T0245259

CAMBRIDGE LIBRARY COLLECTION

Books of enduring scholarly value

Technology

The focus of this series is engineering, broadly construed. It covers technological innovation from a range of periods and cultures, but centres on the technological achievements of the industrial era in the West, particularly in the nineteenth century, as understood by their contemporaries. Infrastructure is one major focus, covering the building of railways and canals, bridges and tunnels, land drainage, the laying of submarine cables, and the construction of docks and lighthouses. Other key topics include developments in industrial and manufacturing fields such as mining technology, the production of iron and steel, the use of steam power, and chemical processes such as photography and textile dyes.

Recollections of Forty Years

The French diplomat and engineer Ferdinand de Lesseps (1805–94) was instrumental in the successful completion of the Suez Canal, which reduced by 3000 miles the distance by sea between Bombay and London. This two-volume memoir, written towards the end of his life and dedicated to his children, was published in this English translation in 1887. In it, De Lesseps describes his experiences in Europe and North Africa. He includes reflections on European and colonial history and politics, a chapter on steam power, and a report on the 1879 Paris conference that led to a controversial and abortive early attempt to build the Panama Canal. Volume 1 focuses on De Lesseps' diplomatic missions to Rome and Madrid in the late 1840s during a period of political and social unrest in Italy, Spain and France, and the early stages of the Suez Canal project.

Recollections of Forty Years

VOLUME 1

FERDINAND DE LESSEPS

CAMBRIDGE UNIVERSITY PRESS

Cambridge, New York, Melbourne, Madrid, Cape Town, Singapore,
São Paolo, Delhi, Dubai, Tokyo, Mexico City

Published in the United States of America by Cambridge University Press, New York

www.cambridge.org
Information on this title: www.cambridge.org/9781108026390

© in this compilation Cambridge University Press 2011

This edition first published 1887
This digitally printed version 2011

ISBN 978-1-108-02639-0 Paperback

RECOLLECTIONS OF FORTY YEARS.

VOL. I.

RECOLLECTIONS OF

FORTY YEARS

BY

FERDINAND DE LESSEPS

TRANSLATED BY C. B. PITMAN

IN TWO VOLUMES

VOL. I.

LONDON: CHAPMAN AND HALL

LIMITED

1887

LONDON :
PRINTED BY J. S. VIRTUE AND CO., LIMITED,
CITY ROAD.

CONTENTS.

VOL. I.

CHAPTER I.

                                                    PAGE
THE MISSION TO ROME   1

CHAPTER II.

EPISODES OF 1848 AT PARIS AND MADRID   .   .   . 119

CHAPTER III.

ROME—SUEZ—PANAMA   129

CHAPTER IV.

THE ORIGIN OF THE SUEZ CANAL   152

THE TRANSLATOR'S PREFACE.

ONE of the greatest of Roman Emperors, when lying upon his death-bed at York, said, "Omnia feci, nihil expedit." And yet, when asked for the watchword of the night, with his dying breath he gave it: "Laboremus." M. de Lesseps, in the course of his long and honoured career, has made the watchword of the dying Emperor his rule of life ; but he is not likely, when his last hour comes, to "look on all the works his hands have wrought and on the labour that he has laboured to do," and find them "vanity and vexation of spirit." For what else was the exclamation of the Roman Emperor but a paraphrase of the Preacher's sermon upon the vanity of all human effort and human enjoyment? In a spiritual sense all this is true enough, no doubt, but the labour of a life mainly devoted to the furtherance of works calculated to benefit others rather than oneself, and to add to the general sum of the welfare of humanity, is not assuredly wasted.

How active and how beneficent a life that of M. de Lesseps has been we most of us know already, though to posterity it must be left to assign the proper place which his name will occupy among the worthies of the nineteenth century. In the meanwhile, the glimpses of the domestic and personal side of his life and character which M. de Lesseps allows us to catch in these volumes cannot fail to be interesting, and some of us may, perhaps, be selfish enough to regret that he has not gone more into detail with regard to this part of his life, even if, to do so, he had been compelled to abridge the account of his diplomatic mission to Rome in 1849 and to omit one or two chapters, such as those upon "Steam" and upon the "Origin and Functions of Consuls," which are of a more general and technical character. I imagine, however, that M. de Lesseps, feeling himself the repository of many secrets, has deemed it best to reserve for some future time the numberless anecdotes which he could, if he were so disposed, relate about the celebrities of every nationality and every profession with whom he has been in contact for upwards of half a century. But the reader of these two volumes will find an abundance of interesting information relating first of all to M. de Lesseps's diplomatic missions to Madrid and Rome, and secondly to the preliminary survey of the Isthmus of Suez, and the intricate negotiations which preceded the actual making of the Canal. As regards the mission to Rome, with which M. de Lesseps commences his "Recollections," I have, while omitting some of the official despatches, the translation of which is not required in order to put the reader in possession of M. de Lesseps's own version of this incident in his career, been careful not to attempt anything like a *précis* of what he says, and this for personal reasons. Having for many years enjoyed the friendship of the late M. Drouyn de Lhuys, who, as Foreign Minister in 1849, entrusted M. de Lesseps with this mission to Rome, I had often heard him speak of it and of the circumstances connected with it. His view was, I need hardly say, diametrically opposite to that expressed here by M. de Lesseps, and I have, therefore, left the latter to tell the story in his own words. I have, however, taken it upon myself to omit two chapters altogether, one being a treatise upon the French Revolution of 1848 by Don Balmès, a Spanish writer, and the other a criticism by M. de Lesseps himself of this author's writings. French readers may possibly be curious to know, even at this remote date, what a Spanish writer has to say about the most deplorable and senseless of the many revolutions which have occurred in their country, but foreigners can scarcely be expected to feel any interest in what is, after all, but the individual expression of opinion by a writer of whom they know nothing upon a subject which has passed quite out of their memory.

To English readers the most interesting, but in some ways the most humiliating, part of the book will

be that in which M. de Lesseps tells at considerable
length the story of how, in face of the stubborn and
unreasoning opposition of Lord Palmerston and other
English Ministers, he carried through his project of
making the Suez Canal. The unflagging energy, the
indomitable perseverance, the never-failing good-
humour with which he met all difficulties and fought
against every kind of obstacle, convey a lesson which
ought not to be thrown away upon the half-hearted
and upon those who are always ready to take no for
an answer. But there is another lesson to be learnt
from the story which M. de Lesseps relates with such
merciless precision. Lord Palmerston opposed the
making of the Suez Canal upon four grounds : first,
because it was impracticable, as he had learnt from
Mr. Robert Stephenson, the engineer, who had not
surveyed more than a small part of the isthmus;
secondly, because, even granting that it could be
made, it would never pay; thirdly, because it was
detrimental to English interests; and fourthly, because it would impair the integrity of the Turkish
Empire, and render Egypt virtually independent of
the Porte. As M. de Lesseps points out, the two
first objections had no force so far as England was
concerned until he came to ask the Government to
subsidise the undertaking; while as to the third, he
triumphantly points out that while the English
Government is denying the right of the Viceroy of
Egypt to make the Canal through the isthmus without

the Sultan's firman, it is urging him to complete the railway from the Mediterranean to the Red Sea without waiting for any such authorisation. M. de Lesseps reminds his readers, too, that England was very glad to use the route through Egypt for sending troops to the relief of India during the mutiny, and he is not less successful in showing that the cutting of the canal could not of itself affect the relations of Egypt to its Suzerain. He shows how, step by step, one country after another rallied to his cause, and how, even in England, public opinion came round to him, the series of meetings which he held throughout the kingdom being unanimous in favour of the scheme. Lord Palmerston, however, remained obdurate, and the English ambassadors at Constantinople—first Lord Stratford de Redcliffe, and then Sir Henry Bulwer—moved heaven and earth to quash the project. But M. de Lesseps, who in this instance at all events showed himself a consummate diplomatist, not only enlisted the active sympathies of several influential persons, such as the Empress of the French, but removed, one by one, the obstacles from his path, and, as we know, brought his work to a triumphant conclusion. Well, I would say that when we come to consider the objections and arguments which are at the present time being urged against other great engineering projects intended to facilitate communication between different parts of the globe, one cannot fail to be struck by their close similarity to those

with which M. de Lesseps was met thirty years ago
when about to commence the Suez Canal. We are told
now that the Panama Canal—in which, as my readers
will all be aware, he is the guiding and controlling
spirit—can never be completed, not at least for a sum
which is, practically speaking, obtainable; and that the
Channel Tunnel between France and England, if not
as impracticable from an engineering and financial
point of view, would be as detrimental to the interests
of England as Lord Palmerston declared the Suez
Canal to be. I do not profess of myself to have sufficient knowledge to speak with authority upon either
of these subjects, though, of course, it can only be in
joke that people talk about the military risk which the
making of a submarine tunnel would involve. But
when one finds how the self-same arguments which
retarded, but did not prevent, the cutting of the Suez
Canal—all of them falsified in the event—are being
served up again by the adversaries of these two projects it is impossible to avoid feeling that a little more
prudence, a little more self-restraint, a little less self-
confidence would not be out of place. Those who declare that the Panama Canal never can be made, and
that the Channel Tunnel never ought to be made, may
be justified by the result, and the arguments which in
the case of the Suez Canal were so utterly falsified
may in these instances prove sound. But, to borrow
a famous phrase, one would be sorry to be " as cock-
sure of anything as they are of everything," and M.

de Lesseps will have increased the debt which the friends of progress and of civilisation owe to him if, by writing the history of his great work in Egypt, he shall have inculcated upon some of us the danger of speaking too dogmatically upon subjects of which our knowledge is necessarily imperfect, and in most cases second or even third-hand.

C. B. PITMAN.

October, 1887.

RECOLLECTIONS OF FORTY YEARS.

CHAPTER I.

THE MISSION TO ROME.

OFFICIAL DOCUMENTS.

WHEN the Austrian army was hanging upon the frontiers of Piedmont, the French National Assembly, using its right of initiative, called upon the Ministry to assume a resolute attitude, and authorised it, should such a step be deemed in the best interests of France, to occupy temporarily some part of the Italian peninsula. A few days after this had been voted, the news of the reverse sustained by the Piedmontese army at Novara reached Paris. The President of the Council laid before the Assembly, upon the 16th of April, 1849, an estimate of £48,000 for the extraordinary expenses which the maintenance of the expeditionary force for three months in the Mediterranean would, it was estimated, entail. The credit was voted, and a force was sent to occupy Civita

Vecchia under the command of General Oudinot, but the Roman Assembly and the Triumvirate declined to receive what they regarded as a hostile force, and ordered the commander at Civita Vecchia to resist to the last extremity. Their order came too late, however, the French troops having already disembarked and entered upon a joint occupation of the place with the Italian troops. General Oudinot, finding that his entry into Rome would meet with a determined resistance, then decreed sterner measures at Civita Vecchia, placing it in a state of siege and disarming the garrison. The prefect of the town having protested against these measures, was cast into prison, and General Oudinot, being anxious to bring matters to a head, marched on Rome, before the walls of which he arrived at the end of April. The result of the combat on the 30th, in which the Roman population spontaneously took so active a part, is well known, and I was able to see for myself that out of every ten Italians whose wounds were being seen to in the hospitals at least eight were natives of Rome. The news of this combat created profound emotion in Paris. The National Assembly, composed of nine hundred members, was very indignant, and showed an inclination not only to upset the Ministry, but to put the Prince President of the Republic upon his trial for treason. The Committee appointed to report upon the situation proposed a resolution calling upon the Government "to take without delay such steps as may be necessary to prevent the Italian expedition from being any longer kept from carrying out the aim assigned to it." The Minister of Foreign Affairs (M. Drouyn de Lhuys) cast all the blame for what had occurred upon General Oudinot, who, he asserted, had received no instructions to attack the Roman Republic. The resolution of the Committee was, however, carried by a majority of 338 to 241, and though this was to a great extent a vote of want of confidence, the Ministry did not resign, but appointed a diplomatic agent, whose mission M. Odilon Barrot, the President of the Council, explained as follows at the sitting of the Assembly on the 9th of May :—

" I assure you that as long as I am in office French arms shall never be used for the restoration of abuses. It is with this feeling, in order to learn from trust-worthy agents the real truth, and also in order to convey to those concerned the faithful and precise expression of the intentions of the Assembly and of the Government in regard to the aim and object of this expedition, that the Government has decided to despatch a man who enjoys our full confidence, whom we have put to the test in very trying circumstances, and who has always served the cause of liberty and humanity. M. de Lesseps, to give you his name, has been sent, and we have specially instructed him to place himself in immediate communication with the Government and to keep us informed day by day of whatever may happen. We have further impressed upon him that he is to employ his utmost influence so that our intervention may secure genuine and real guarantees of liberty for the Roman States."

I should have said that the Minister of Foreign Affairs had sent for me on the morning of the 8th, after the day and night sitting of the Assembly, at both of which I had been present, and had asked me if I was disposed to undertake a very important mission for which the Ministry, at the Cabinet Council just held, had selected me.

I replied that, as I had been deemed worthy of so high a mark of confidence, I felt it my duty frankly to declare that if the Government had not, at the outset, been animated by an open and resolute policy, it would have been much better not to have compromised us by sending an expedition to Civita Vecchia. However, I added, the point now is to repair the mischief done by the affair of April 30th and not to fall into the same blunder again. I also said that I should be ready to start in two hours if necessary, and I promised to leave no stone unturned to arrive at the aim indicated in the vote of the previous day. M. Drouyn de Lhuys congratulated me upon my readiness, and added that the manner in which I expressed myself went far to show that the Government had made a judicious selection. While I was with him he sent for the chief clerk of the political department, M. de Viel-Castel, and requested him to draw up instructions which would leave me such latitude and initiative that my political action would not be hampered either by the general entrusted with the military operations or by orders which would clash with unforeseen events that might have occurred since the 30th of April. He also advised me to get several copies of the *Moniteur* (then the French official journal, which contained the reports of the debates in the Assembly) and hand over one to General Oudinot as soon as I arrived, being of opinion that it was upon the decision of the Chamber that we should base our course of action.

The text of these instructions was as follows:—

" The events which have marked the first steps of the French expedition sent to Civita Vecchia being calculated to complicate a question which at first seemed a very simple one, the Government has come to the conclusion that it is advisable to appoint, in addition to the military commander of the forces despatched to Italy, a diplomatic agent, who, devoting himself exclusively to the negotiations and relations to be established with the Roman authorities and inhabitants, will be able to give these grave matters the close attention and anxious care which they require. Your tried zeal, your experience, and the conciliatory disposition which you have more than once displayed in the course of your career, have induced the Government to select you for this delicate mission. I have explained to you the present state of the question upon the solution of which you are about to enter. The object which we have in view is at once to deliver the States of the Church from the anarchy which prevails in them, and to ensure that the re-establishment of a regular power is not darkened, not to say imperilled, in future by reactionary fury. Any step which, in presence of the intervention exercised by other Powers animated by less moderate views, will give more scope to our special and direct influence, will naturally make the object which I have pointed out to you more easy of attainment. You will therefore concentrate all your efforts upon bringing about such a result with as little delay as possible; but in the efforts which you will make towards this end, there are two risks to be guarded against, as I will point out to you. You must be careful to avoid allowing the men at present invested with power in the Roman States to suppose that we regard them as a regular Government, for that would give them a moral force in which they are at present lacking. It will be desirable, in the partial arrangements which you may conclude with them, to avoid using any expression or making any stipulation which may be likely to excite the susceptibilities of the Holy See or of the Gaëta conference, which is only too ready to assume that we are inclined to attach no value to the authority and interests of the Court of Rome. Upon the ground on which you will be standing, and with the men with whom you will have to deal, the ques-

tion of form is almost as important as that of principle.
These are the only instructions which I can give you,
for, in order to render them more precise and more
detailed, it would be necessary to have before me
information, not yet forthcoming, of what has happened in the Roman States during the last few days.
Your upright and enlightened judgment will inspire
you according to circumstances. But in any case you
will confer with MM. d'Harcourt and de Rayneval
in reference to all matters which do not call for an
immediate solution. It will be superfluous for me to
engage you to maintain close and confidential relations with General Oudinot, this being absolutely
essential to the success of the enterprise which you
are called to work out in common.

" Drouyn de Lhuys."

M. Drouyn de Lhuys himself read me these instructions, and, dwelling at the first passage which
authorised me " to devote myself exclusively to the
negotiations and the relations to be established with
the Roman authorities and inhabitants," he pointed
out to me that this left me a very large share of
authority independent of the general in command.
He dwelt also upon the final paragraph which left
me full latitude in presence of unforeseen difficulties
or incidents.

With regard to the passage relating to concerting
with MM. d'Harcourt and de Rayneval, I pointed

out that such a concert was impossible, as their mission and mine had a quite different, not to say contrary, principle. The answer was: " Simply send them duplicates of your despatches."

I was still with him when a message from the Prince President summoned me to the Elysée, where I had already been in the morning; and M. Drouyn de Lhuys asked me to come and let him know what passed between us.

The Prince told me that he had carefully considered the object of my mission, and that one point, about which he was afraid that he had not spoken to me, gave him great concern. This was the attitude of our troops in the event of an armed intervention of the Austrians and Neapolitans, whose action must at all costs be prevented from being brought into common with ours. He gave me, in connection with this, a letter for General Oudinot, and asked to see my instructions, which he thought rather ambiguous, and not sufficiently explicit. He informed me that he intended sending to Rome General Vaillant, who would be instructed to come to an understanding with me, and who would replace General Oudinot if the latter did not hit it off with us, or assume command of the engineering operations if the siege of Rome should be renewed. He added that I should do well, if the opportunity occurred, to call attention to the fact that in 1831 he had already taken part against the Temporal Power when he was before

Rome in the company of his elder brother, who died during the insurrection.

Upon returning to see M. Drouyn de Lhuys, I was careful not to confide this matter to him, nor did I make any use of it while in Rome, so as not to excite public feeling unnecessarily. But when I repeated to him the Prince's observation about a foreign intervention in the Roman States, he asked me how I interpreted the expression "at all costs" as applied to preventing anything like a common action with the Austrians and Neapolitans. I told him that it was for him to settle that with the President and write to me ; but that, until I heard further, I should interpret it in the widest sense. M. Drouyn de Lhuys's salon being then full of visitors, as it was his regular reception-day, I took leave of him and was soon travelling in a post-chaise to Toulon, where telegraphic orders had been sent for a man-of-war to be got ready for me, upon which M. Drouyn de Lhuys had given leave for Signor Accursi, a friend of Mazzini and Minister for Home Affairs of the Roman Republic, also to travel. M. Drouyn de Lhuys had suggested that Signor Accursi should accompany me to Toulon ; but I pointed out that this might be compromising.

Before embarking I received two despatches from the Ministry of Foreign Affairs for MM. d'Harcourt and de Rayneval and for myself, which I give in their entirety.

" *The Minister of Foreign Affairs to MM. de Rayneval
and d'Harcourt.*

" PARIS, *May* 9, 1849.

"What pains us even more deeply than the mistrust which is still shown us at Gaëta, but which time will eventually dissipate, is the nature of the influences which evidently prevail in the councils of the Holy See. The nearer we seem to the *dénouement*, the more clearly come out dangerous propensities which are for the moment disguised beneath more or less specious pretexts. In order to avoid making any set declaration as to the intentions of the Holy Father, his advisers say how inconvenient it would be for them to have his hands tied. There might be something in this objection if it were necessary to settle in detail the basis of a fresh *régime ;* but when all we ask is what course it is intended to follow, once the authority of the Holy See is re-established, it is hard to understand why the Holy See should wrap itself in impenetrable silence, unless there is a hidden resolve to return simply to all the abuses of the ancient *régime.*

"We are told that there are certain reactionary tendencies among the populations which must be treated tenderly, and of which we have not taken sufficient account. If these tendencies had the great force which is attributed to them, would it not be advisable to assume without delay an attitude which would at some future time place the Holy See in a position to combat them ? Is it supposed, moreover, that there is no need to reassure that numerous portion of the Roman population which, while detesting the rule of anarchy, dreads almost as much the return of one who has left so melancholy a mark upon the reign of Gregory XVI. ; of a *régime* which at the death of that Pontiff had rendered a change of that system absolutely necessary, and which, by provoking a vigorous reaction, has done far more to bring about the misfortunes of these recent times than the hurried introduction of certain reforms which were not, perhaps, sufficiently thought out. The men of whom I speak, and who, if I am not mistaken, comprise nearly the whole of the well-to-do and enlightened classes, would gladly rally now to any combination which offered them guarantees of good order, security, and sound administration; but how can they be otherwise than uneasy when they see that not a word is said as to the future ? and are they not justified in fearing that there is a design to annul all the concessions due to the generosity of Pius IX., including the secularisation of public functions, the prime and essential basis, without which any reform attempted in the States of the Church can be but illusory ?

" I will say no more upon this subject. You are aware of the painful reflections which it has forced upon me, and you have done your best to bring those who obstinately refuse to recognise the truth of these reflections to take a more accurate view. In refusing to allow the Holy Father to reassure the public mind by explanations and promises, they have probably contributed to intensify the unexpected resistance which our expedition has encountered. They trust to the Pope being brought back to his States by foreign aid, but have they taken any thought of the future which they are preparing for him by urging him to take this deplorable course? Will the excuse of an ill-omened success, of an attempt at reform made in the most deplorable conditions, have more weight than all the arguments of reason, backed up by so many examples borrowed from the history of these recent times?

" Be this as it may, what we are now doing for the pacification of the States of the Church, the sacrifices which an undertaking of this kind entails upon us, and the moral responsibility which it imposes upon us, unquestionably justify us in urging that a line of conduct which would so intensify that responsibility shall not be persisted in.

" The desire which we thus express does not, moreover, go beyond what might legitimately be expected. We only ask for what has up to the present time been promised us without any difficulty. It is the realisation of a line of conduct which up till the other day did not seem to be at all questioned. We were repeatedly being told that a return to the ancient *régime* was impossible ; that the present state of

men's minds and the general situation of Europe were
imcompatible with it; and the most that was hinted
was that it would perhaps be prudent to make some
slight changes in the constitutional statute granted
by Pius IX. The necessity, the expediency of such
changes may be taken into consideration when order
and peace have been re-established; but I must add
that we do not admit that this statute itself can in
future be regarded as null and void. The respect we
entertain for the Holy Father prevents us from
admitting that the institutions which he granted to
his people have been completely annulled by the
deplorable events which have occurred in Rome since
last November. The idea that the *régime* anterior
to 1846 would be revived in Rome never entered into
our minds or calculations. We acted under the
influence of quite an opposite conviction.

"We still hope that we were not mistaken. We
do not wish to attach too much importance to a few
words hastily uttered, perhaps in a moment of excitement, but interests of too high an order are at stake
for me to await explanations which would perhaps
dissipate our anxieties before instructing you to make
to the Cardinal Secretary of State, to the Holy Father
himself, and, if you think it well, to the members of
the Conference, representations the urgency of which
must of course be in proportion to the gravity of the
dangers which they are designed to avert. They will
understand that, in the position we hold, we have

serious duties to perform, and these duties we are resolved to fulfil.

"Do not lose an hour in letting me know what answer you have received to the pressing advice of which you will find the text inclosed. It is important for us to know what we have to expect.

"DROUYN DE LHUYS."

The Minister of Foreign Affairs to M. Ferdinand de Lesseps.

"PARIS, *May* 10, 1849.

"We have learnt that General Oudinot had thought it incumbent to request a commissioner sent, in the name of the Holy Father, to Civita Vecchia, not to prolong his stay in that town, where his presence produced a bad effect. The Government of the Republic quite approves of this step, dictated by a sound comprehension of the necessities of the situation and of what the safety of our army demands.

"Until the object of our expedition has been attained we cannot allow centres of authority to be organised outside our influence upon the territory we occupy, which might, even if unintentionally, go counter to our action and compromise its success. It is with this in view that you will make use of the powers which have been confided to you. No one better than yourself will know how to use these powers with the necessary degree of firmness and at the same time with such consideration as to spare susceptibilities, of which it is extremely desirable to take all possible account.

" DROUYN DE LHUYS."

At the same time, the Minister of Foreign Affairs, under the impression of the vote of May 7th and of the refusal of the Gaëta Cabinet to assist our enterprise, and to promise the inhabitants of Rome liberal institutions, sent to General Oudinot the following telegram intended to precede me, in the event of my journey to Italy being delayed :—

" PARIS, *May* 10, 10 A.M.

"Inform the Romans that we do not intend to join with the Neapolitans against them. Follow up the negotiations in the sense indicated by your instructions. Reinforcements are being sent to you, await their coming. Endeavour to enter Rome with the assent of the inhabitants, or if you are compelled to attack, do so with the most absolute certainty of success."

I reached the head-quarters at Castel de Guido at one in the morning. Being at once taken to see the General, who was ill in bed, as a result of his repulse before the walls of Rome, I read him from the *Moniteur*, a copy of which I left with him, the report of the debate of the 7th in the National Assembly, and communicated my instructions to him. He promised me his help in the accomplishment of my mission. As he could not use his pen, I wrote, under his dictation, to Count de Ludolf, Minister of Foreign Affairs to the King of Italy, who was encamped with his escort and Neapolitan army upon the other side of Rome, to let him know how matters stood, enclosing with my letter a copy of the *Moniteur*.

As my arrival was to modify the operations already begun, the General lost no time in sending out orderlies in different directions, so that any offensive movements which might hamper my negotiations should not be carried out.

As soon as it was daylight, I went into Rome, accompanied by M. de la Tour d'Auvergne, Secretary of Legation. We had some difficulty in obtaining admission, and it was necessary to make a partial circuit of the walls, as several of the gates were barricaded. All along the road were posts upon which were inscribed in large letters the clause of our constitution which forbids any attack upon a foreign nationality. Some of the sentinels upon the ramparts levelled their rifles at us, but my servant, who was sitting beside the driver of our brougham, flourished a white handkerchief, and the rifles were at once lowered. At last I saw a gate open, and a young officer, Colonel Medici, who recognised me, came forward and offered me his services, saying that the city of Rome would be glad to hear of my arrival.

He had me accompanied by a detachment of his men to the Via Condotti, where I alighted at the Hotel d'Allemagne, thinking it advisable not to go just yet to the French Embassy.

After having several visits paid me, among the visitors being Charles Bonaparte (Prince de Canino), President of the Assembly, I forthwith wrote as under to the general in command :—

"Having regard to the expectant attitude in which we are placed, it seems to me of the utmost importance to avoid any sort of engagement. I find a whole city in arms, with the population apparently bent on resistance, while, without any exaggeration, there are 25,000 men ready to fight. If we entered Rome by force, not only should we have to do so over the bodies of a certain number of foreign adventurers, but we should have to strike down a great many shopkeepers and young men of good family, representatives of the classes which defend social order in Paris. We must, therefore, take account of this situation, not act precipitately or implicate our Government in anything opposed to the object which it had in view at the beginning of the expedition— an object which it has just declared anew—or to the wishes of the National Assembly. I should, therefore, hold myself much to blame if I did not use my best efforts to induce you to suspend all acts of hostility or any demonstrations likely to bring them about until I have seen you, and been able to give you an

account of what I have seen. You are of my opinion,
I know. At the same time, I shall declare that our
soldiers will not budge an inch. Your attitude and
your kindly disposition cannot fail to facilitate an
honourable arrangement. We are strong, we can
afford to wait."

I came to a verbal understanding with General
Oudinot and the Roman authorities as to a suspension
of hostilities.

Having obtained this result, I endeavoured to form
a correct estimate of the situation and of the difficulties by which I might expect to be confronted. I
was not long in discovering that in Rome I should
have to face the prejudices of a population still very
irritated by the events of April 30th; the impossibility in which we found ourselves placed of officially
recognising the Roman Republic or even of promising
the maintenance of a government which esteemed
itself to be as legitimate as our own; and the blind-
ness of certain influential persons who were relying
for the triumph of their cause upon a revolutionary
movement in Paris, just as many French politicians,
even in the ministerial party, believed in the
existence of a *moderate Roman party*, which had
promised to open us the gates of Rome on the 30th
of April, and would be more fortunate if we attacked
the city again.

Upon the other hand, I had remarked that the
impatience of several generals, the desire to make

amends for a personal check, the constant instigations of persons interested in a renewal of hostilities, and the echo of the unenlightened advice by which the Holy Father was guided, would raise up for me at the French head-quarters obstacles which, if less imminent, would perhaps be more persistent than those which I had just surmounted in Rome.

DIRECTION POLITIQUE. No. 1.

First Despatch to M. Drouyn de Lhuys.

"ROME, *May* 16, 1849.

" Monsieur le Ministre,—I informed you yesterday by telegraph that, after having come to an understanding with General Oudinot, I should start for Rome accompanied by M. de la Tour d'Auvergne, in order to ascertain for myself the real sentiments of the Roman population and supply you with an exact account of the information which I obtained. I subjoin you a copy of the letter which I wrote the same day to General Oudinot from Rome. M. de la Tour d'Auvergne quite shares my ideas. M. de Gérando, a man of good sense, whom I had heard highly spoken of at the Ministry of Foreign Affairs before I left Paris, confirms my opinion as to the resistance which would be offered us being a very general one. Not that I doubt the ultimate success of our arms, but it would only be reached through

a sea of blood, and that is what neither you nor I
desire.

"I had not been long in Rome before the Triumvirs
expressed their wish to see me. When I called upon
them I informed them that I had been sent by my
Government to ascertain and to speak the truth
as to the state of public feeling in Rome since the
events of April 30th; that our object was to employ
all the means compatible with our dignity and
military honour to prevent a deplorable struggle
between the French and the Romans; that, after what
I had seen and should communicate to General
Oudinet, I hoped soon to be able to announce that
all hostile acts or demonstrations upon the part of the
French army against Rome would come to an end.

"This morning I sent M. de la Tour d'Auvergne to
head-quarters, and he informed the General of what
I had done, and brought back with him the latter's
promise not to hamper my action by any hostile
demonstration. I am therefore in a position to
promise, upon behalf of the General in command as
well as of myself, that hostilities would be suspended,
and to show myself ready to enter into negotiations.
I have confidently suggested that the National
Assembly should send a deputation selected from its
midst to head-quarters to negotiate, and should ask me
to accompany it. I am in hopes that this suggestion
will be adopted; and I have already ascertained that
the Triumvirs, the President, and several deputies of

the Assembly, and many other persons of influence
over the inhabitants, are favourable to it. The result
seems well assured, and it does not compromise us in
any way, as the object is to enable us to negotiate
with the executive of a Government which we cannot
recognise officially. It has been arrived at after
exertions which have not left me a minute's leisure.
When I arrive with a deputation from the Assembly
at head-quarters, it will be the time to come to some
arrangement. I have just drawn up a scheme, of
which a copy is annexed herewith. I shall go and
discuss the basis of it to-morrow morning with the
General, and probably with M. d'Harcourt, whose
arrival is announced as imminent. You will see for
yourself whether he conciliates the very complicated
interests which we have to study, whether he reserves
for the Government of the Republic full liberty to
pursue, according as its interests and fresh circumstances may dictate, a clear and resolute course of
policy.

" A column of 12,000 men, infantry, cavalry, and
artillery, under the command of Garibaldi, left at
five o'clock this afternoon to attack the Neapolitans.

" FERDINAND DE LESSEPS.

" P.S.—I have paid a visit, in company of M. de la
Tour d'Auvergne, to two hospitals in which twenty-
six French soldiers, wounded in the engagement of
April 30th, are under treatment. I promised them

that they should rejoin their comrades as soon as they were cured. They could not possibly be better cared for, as Roman ladies of the highest families are tending them day and night, having taken up their residence in the hospitals, the Princess Beljioso at their head."

Second Despatch to the Minister of Foreign Affairs,
Paris.

"Rome, *May* 18, 1849.

"Monsieur le Ministre,—A conference has been held at the head-quarters of the French army between General Oudinot, M. d'Harcourt, and myself. I read and commended my first despatch. My preliminary measures were approved, as well as the scheme destined to form a basis for negotiations. This scheme will undoubtedly undergo modifications of detail which will not alter its main principles.

"Upon my return to Rome I learnt that the Assembly had unanimously decided that a committee of three members should be selected, and the members chosen were MM. Sturbinetti, Audinot (Bologna), and Cernuschi (Milan). This latter, who would have been a very desirable choice, as indeed were the two others, would not accept out of delicacy, and it occurred to me that it would be better that the deputation should consist of Italians who were natives of the Roman States. His successor was chosen to-day, Signor Agostini.

" Previous to the sitting at which yesterday's resolution was carried, I had several visitors in my room,
among others M. Charles Bonaparte, who was to take
the chair. An attempt was made to draw a distinction
between my intentions and those of General Oudinot
and the French Government. I was asked what
could be done to destroy the prejudices which existed
in this respect among the Roman population. I then
told them that nothing could be easier, as you had
just written to me under date of the 10th, signifying
your approval of the conduct of General Oudinot, who
had thought it best to expel from Civita Vecchia an
envoy of the Pope, whose presence was calculated to
produce a bad effect and hamper our action. I need
scarcely assure you that I do not say a word more
than is necessary to extricate us from one of the most
difficult positions in which we have been for a long
time placed ; that in all other respects I am very
reserved in my relations with every one ; and that if
I listen to men of all nations, all sorts, and all
parties, who come to see me as early as five in the
morning and as late as midnight, giving them all a
cordial welcome, it is in order to accomplish to the
best of my ability the mission you have entrusted
me with.

" To-morrow probably will begin the negotiations ;
I am starting for head-quarters in order to concert
about them with the General, with whom I am on
such terms as might be expected from his patriotism

and from the loyalty of his character.   I take care to
arrange with him in respect of everything which bears
upon our common instructions.

" M. de Forbin-Janson will convey this despatch
and the preceding one.   M. d'Harcourt has authorised
him to proceed to Paris in order that he may tell you
what he has seen and what the present condition of
Rome is, as his information is likely to be of use.

" I begged Signor Mazzini to hand me a note
explaining his views as to the present situation of
Rome, and he readily acceded to my request.   I have
the honour to forward you a copy of his letter, which
you will, I feel sure, consider a very remarkable one.

" FERDINAND DE LESSEPS."

Annex to Despatch No. 2.

"Sir,—You ask me for a few notes upon the present condition of the Roman Republic, and I will write
them for you with that frankness which has for twenty
years been the invariable rule of my political life.   We
have nothing to conceal or to disguise.   We have of
late been singularly vilified in Europe, but we have
always said to those to whom we have been so calum-
niously denounced, ' Come   and   see   for yourself.'
You are here, and can therefore verify for yourselves
how far these accusations are true.   Your mission can
be carried out with full and complete freedom.   We
have saluted it with joy, for it is our best safeguard.

"France doubtless does not question our right to govern ourselves in our own way, the right to draw, so to speak, from the entrails of the country the idea which regulates its existence, and to make of it the basis of our institutions. All that France can say to us is, 'In recognising your independence, it is the free and spontaneous wish of the majority which I desire to recognise. United to the European Powers, and being anxious for peace, if it were true that a minority among you sought to oppose the national tendencies, if it were true that the present form of your government was only the capricious fancy of a faction, substituting itself for the common aspirations, I could not look on with indifference while the peace of Europe was being constantly endangered by the turbulent scenes of anarchy which necessarily characterise the reign of a faction.' We recognise that France has this right, for we believe in the solidarity of nations for good. But we maintain that if ever there was a Government springing from the wishes of the majority of the nation, and maintained in power by it, that Government is our own.

"The Republic has been implanted in our midst by the will of an Assembly elected by universal suffrage. It has everywhere been accepted with enthusiasm; nowhere has it encountered the least opposition.

"And you must remember that never was opposition so easy, so devoid of danger, I may even add so provoked, not by its acts, but by the exceptionally unfavourable circumstances in which the Government started upon its career.

"It followed upon a long anarchy inherent in the organisation of the Government which preceded it. The agitations which are inseparable from all great transformations, and which were at the same time fomented by the crisis of the Italian question and by the efforts of the retrograde party, had thrown it into a state of feverish excitement which rendered it accessible to any bold attempt, to any appeal to interests and passions. We had no army, no repressive powers. As a consequence of previous waste our finances were impoverished, not to say exhausted. The religious question, in the hands of able and interested persons, was available as a pretext to work upon the feelings of a population of gentle instincts and peaceful aspirations, but not very enlightened.

"And yet no sooner had the Republican principle been proclaimed than order prevailed. The Papal Government does not tell us anything about its insurrections: there has not been one under the Republic. The assassination of M. Rossi—a deplorable but an isolated event, an individual act reproved and condemned by us all, provoked it may be by an imprudent attitude, and the source of which has never been traced —was followed by the most perfect order.

"The financial crisis reached its highest pitch; there was a brief period during which the paper money of the Republic was reduced, through dishonourable manœuvres, to a discount of 41 or 42 per cent. The attitude of the Italian and European Governments became more and more hostile. These difficulties, as well as their material isolation, the people endured with the utmost calm, having faith in the future which would be evolved out of the new principle proclaimed.

"By means of obscure threats, but more especially owing to unfamiliarity with political habits, a certain number of electors had abstained from contributing to the formation of the Assembly, and this seemed to weaken the expression of the national will. But a second and very vital and characteristic fact triumph-antly refuted any doubt there might have been upon this score. Shortly before the formal institution of the Triumvirate there was a fresh election of the munici-palities. The vote was a large one, and though the municipal element is always the most conservative one in the State, so much so that we were for a time afraid that the elections would show retrograde tendencies, the municipalities selected this very moment to give in their spontaneous adhesion to the new form of govern-ment. During the first fortnight of this month we received, in addition to those of the clubs and the commanders of the National Guard, the addresses of all the municipalities with two or three exceptions. I have had the honour to send you the list of them. They all proclaim explicitly their devotion to the Re-public, and a profound conviction that the two powers united under one head are incompatible with each

other.   This, I repeat, constitutes a decisive fact.   It
is a second legal proof completing the first in the most
absolute manner, and testifying to our rights.

"At the present time, amid the crisis which
prevails, in presence of the French, Austrian, and
Neapolitan invasion, our finances are improved and
our credit restored; our paper is discounted at 12 per
cent. ; our army is increasing each day, and there are
whole masses of men ready to rise and reinforce it.
You have seen Rome, and you know how heroic a
struggle Bologna is carrying on.   I write this at
night amid the most profound calm.   The garrison
left the city last evening, and before the fresh troops
could come in, at midnight, our gates and barricades
were, by means of a simple password handed on
from mouth to mouth, guarded without noise or
ostentation by the people in arms.   There is in the
heart of this people one resolute determination, and
that is the downfall of the temporal power invested in
the Pope, the hatred of priestly government under
whatever attenuated or indirect form it may present
itself.   I say hatred, not of individuals, but of the
government.   Towards individuals our people have,
thank God! ever since the foundation of the Republic
shown themselves generous, but the very idea of a
clerical government, of a pontiff-king, makes them
shudder.   They will fight to the death against any
scheme of restoration, and show themselves schismatic
to the last rather than endure it.

" When the two questions were submitted to the Assembly, there were a few timid members who thought that the proclamation of the Republic might be premature and dangerous in the present state of Europe, but not one to vote against the downfall of the Papacy, right and left uniting to declare that the temporal power of the Pope was for ever abolished.

" With such a people what can be done ? Is there a single free government which, without committing a crime and contradicting its essence, can assume the right to impose upon it a return to the past ?

" Remember that a return to the past means neither more nor less than organised disorder, a renewal of the struggle of secret societies, the uprising of anarchy in the heart of Italy, the inoculation of vengeance into a people which is only desirous of forgetting, a brand of discord permanently implanted in the midst of Europe, the programme of the extreme parties supplanting the orderly Republican Government of which we are now the organs.

" This surely cannot be desired by France, by her Government, by the nephew of Napoleon; especially in the presence of the double invasion of the Neapolitans and Austrians.

" An attitude of hostility against us just now would recall in some measure the hideous concert of 1772 against Poland. Such a design would, moreover, be impossible to realise, for the flag hauled down by the will of the people could only be raised aloft again over a mountain of corpses and the ruins of our cities.

" I shall have the honour of submitting some other considerations to you to-morrow or the day after.

" Your very devoted

" JOSEPH MAZZINI."

Signor Mazzini to M. de Lesseps.

" [Private.]

" ROME, *May* 17, 1849.

" The messengers carrying the ordinary correspondence, who go out by the Angelica gates, have just been driven back by the French, at the orders of the General. What does a cessation of hostilities mean if we are still to be hampered and embarrassed in keeping up communications with the provinces and in preparing our means of defence against the Austrians and the Neapolitans? The only effect it can produce upon our populations is to induce the belief that the truce is, so far as it regards ourselves, a word void of meaning! This state of things cannot last. Remember that our territory is invaded, and that we must defend ourselves. The messengers were stopped at the Acqua Traversa bridge. See if you can set this right. I know the country, and I am certain that all negotiation will be impossible if this state of things lasts.

" JOSEPH MAZZINI."

M. de Lesseps to Signor Mazzini.

"Rome, *May* 18, 10 A.M.

"I received your letter upon my return late last night. The matter of the messengers shall be arranged at once.

"It might be inferred from something said in the Chamber yesterday that an attempt would be made to distinguish between the conduct and the intentions of my Government. I think it fair to inform you that if the Powers with which we are about to treat entertain any idea of this kind, or if a language which would be the consequence of it should be made use of, either against the President of the Republic, the Ministers who sent me to Rome, or the honourable General Oudinot, all negotiations would be at once broken off.

"My Government has been charged with having some afterthought. If this were the case I should not have been entrusted with a loyal and humane mission which I intend to fulfil to the very last, and in connection with which I have already found that I can count upon your able co-operation. I do not doubt but what I shall succeed, inasmuch as the result which we were endeavouring to arrive at is one which will bear the light of day.

"I have sent your note on to M. Drouyn de Lhuys. I thank you for it.

"F. DE LESSEPS.

"P.S.—I authorise you to make what use you may think proper of this letter."

M. de Lesseps to the Minister of Foreign Affairs.

"Rome, *May* 22, 1849.

"M. de la Tour d'Auvergne, whom I am sending to Paris, and whom I beg you to send back to me at once, will give you the information which I consider indispensable in the present juncture for the further-ance of our policy. It is impossible for me to furnish you with it by letter, for at this moment my *rôle* is altogether an active one, and leaves little time for despatches. The documents annexed form the main elements of the information which M. de la Tour d'Auvergne will be well able to lay before you. I have the utmost confidence in him, and could not be better seconded than I am by him. The documents which he will submit to you are :—

"1st. A draft of arrangement as modified after the discussions which have taken place with the Commis-sioners of the Roman Assembly.

"2nd. Note of explanation handed to the Roman Commissioners and the equivalent of the *procès-verbal* of the conference.

"3rd. Letter addressed to me, on the 19th, by the members of the Triumvirate.

"4th. Copy of private letters exchanged on the 21st inst. by General Oudinot and myself.

"5th. Extracts of a correspondence with Signor Mazzini.

"6th. Letter from Commander Espivent, General

Oudinot's head aide-de-camp, and letters from General Oudinot with reference to an ambulance waggon presented to the Roman hospitals in recognition of the attentions shown to our soldiers who were wounded on the 30th of April.

"7th. Note addressed to the Triumvirate.

"8th. Collective declaration communicated to the Roman Assembly and the Triumvirate.

"9th. Reply of the Triumvirate.

"It will be gathered from the two last documents that the course deemed most in harmony with our interests is to allow the Roman population, which seems favourable to our proposed settlement, to manifest its sentiments in such a manner as to bring the men who are at its head to a true appreciation of their interests. I have thought it right to urge upon General Oudinot that the suspension of hostilities should be prolonged, so that the French Government may have sufficient time to see its way and decide upon its course after receiving information of a trustworthy character.

"But whatever the solution may be, I do not think that our expeditionary force is strong enough, taking into consideration the increase of the defensive works and the general arming of the population. After having carefully discussed the matter with the general in command, and after having gone over the city with his first aide-de-camp, M. d'Espivent, who has been here with me for the last two days, I am convinced that it is necessary to send off from twenty to five-and-twenty thousand men from Toulon and Marseilles. This will be none too many. In the event of an understanding being come to with Rome, and our troops entering the city as allies, it would be desirable that our troops should be on the road before the arrangement, which I see no necessity for hurrying on, is completed. If we sent for reinforcements after the occupation of Rome, to overcome any fresh difficulties which might arise, this step might, in the midst of a population which would have received us as friends, tend to aggravate our difficulties. If we are very strong before any definite step is taken we shall terminate matters far more expeditiously and at less cost, and we should be able to send our troops home afterwards far more quickly. We must not lose sight of the fact that the increase and concentration of the French forces at Civita Vecchia and Rome will not weaken us internally, for when once our flag is firmly implanted in Italy we shall have no more risings to put down, and in any event it must be borne in mind that we have now to do not with the soldiers of the Pope, but with the Roman soldiers.

"I am of opinion that General Oudinot should be kept where he is. Whatever you do, do not send him any more siege material. What he wants is a reinforcement of troops, and if he gets them Austria will hesitate to attack us, whereas with fresh siege material it will seem as if we are determined to annihilate Rome, to which I will not in any circumstance whatever lend a hand. And if the intentions of the Government should happen not to be what I believe them to be, I do not hesitate to ask you to recall me, for if I had not my liberty of conduct and was not free to act as circumstances might dictate in the midst of this very complicated crisis my position would be untenable. I shall continue therefore to act without hesitation, and in spite of all material and personal obstacles, until M. de la Tour d'Auvergne has given you by word of mouth the details which it is impossible for me to furnish you with by letter, and until you have informed me by telegraph, yes or no, whether I am in agreement with you.

"It is of set purpose that I have altered clause 3 of the draft of arrangement. I have endeavoured to reduce it to its most simple expression by eliminating all that is not urgent, and by avoiding the two dangers which were pointed out to me—that of formally recognising the Roman Republic, and that of exciting the susceptibilities of Gaëta by alluding to the conflict between the Holy Father and the liberties of Rome. I came to the conclusion, after mature consideration, that by maintaining this clause as it stood we should at once shut the door upon any attempt at conciliation.
"F. DE LESSEPS.

"P.S.—I have just come in from head-quarters. I have read this despatch to M. d'Harcourt. He protests against the inaction of the army, without, however, setting himself against the carrying out of my

advice, accepted by General Oudinot. I am still
going forward with it. You will see which of us is
right. If not agreed upon these points, we are on
very cordial terms."

Ever since the 16th inst. I had been agreed with
General Oudinot as to the drawing up of the following
project, opposed by M. d'Harcourt, which I forwarded
to the Ministry, with an intimation that it would
undoubtedly be modified in some particulars:—

" Clause I.—No restriction shall be in future placed
by the French army upon the liberty of communications between Rome and the rest of the Roman States.

" Clause II.—Rome shall treat the French army as
a friendly force.

" Clause III.—The present executive power shall
resign and be replaced by a provisional government
composed of Roman citizens, and appointed by the
Roman National Assembly, until the inhabitants,
having been called upon to express their wishes, the
Senate shall have decided as to the form of government by which they are to be ruled, and as to the
guarantees to be given in favour of the Catholic religion and the Papacy."

I soon found, after a preliminary conference with
the Roman authorities, that this question could not
even be discussed without awkward consequences, and
that Clause III., relating to the resignation of the
executive power, would lead to interminable discus-

sion. Moreover, such a clause, much as it was desired by the general in command, did not appear to me to form part of my instructions, or to be adumbrated by the speeches of M. Drouyn de Lhuys, who, at the sitting of the 7th inst., had defied the Opposition to find any proof of the Roman Government having been called upon to resign its powers. The first proclamation of General Oudinot, drawn up, as has been shown, by the Minister himself, was, moreover, very explicit upon this point, as it ran: " We will concert with the existing authorities in order that our momentary occupation may not in any way embarrass you." I found, moreover, after having well informed myself as to the actual state of public feeling, that as my mission was a special one and designed solely to effect a conciliation between the French army and the population of Rome, it would be prudent to reserve in all their integrity the questions relating to the Holy See, and not to allow his sacred person or Catholicism to be dragged into a public discussion, the tone of which it was impossible to foretell. It seemed therefore better policy to limit these discussions to the subject of a partial arrangement, as an indispensable preliminary to the general negotiations which might be carried on later between the different Governments. A fresh wording was accordingly agreed upon by General Oudinot and myself. The three commissioners elected by the Roman Constituent Assembly were appointed to discuss it with me. They called upon me, but as they said that the Assembly had given them no powers beyond those of hearing what I had to say and referring back to it, I did not deem it fitting to open up any conferences with them at head-quarters. They were merely charged with the duty of submitting to the Assembly the three following proposals:—

" Clause I.—The Roman States request the fraternal protection of the French Republic.

" Clause II.—The Roman populations have full right to decide for themselves upon the form of government.

" Clause III.—Rome will welcome the French army as a friendly force. The French troops will assist in maintaining order in the city. The Roman authorities will act in accordance with their legal functions."

I had made my mind up quite clearly in changing the third clause of this draft of agreement. I had sought to reduce it to its most simple expression, and to avoid the difficulties already referred to.* The first draft made no reference to the occupation of Rome by the French army; the second made special mention of it, in deference to the views expressed by General Oudinot and M. d'Harcourt. Although my own opinion was contrary to the military occupation of Rome by the French army, for reasons which I will give presently, I did not like to refuse to demand it, in spite of my fear that it would not be granted, not wishing to begin by separating my views from those

* Note of the Translator. See p. 14.

of persons with whom I was desirous of acting in concert. It was on this account that I was led to propose that the military service of the capital should be performed by our troops conjointly with those of Rome. Nevertheless, we had already before us the example of Civita Vecchia, which was by no means calculated to encourage a continuance of the same system.

The Triumvirate informed me, in a note dated the 19th, that our proposals could not be accepted, as they were not considered to offer a sufficient guarantee in favour of the liberties and independence of the Roman States, and because the military occupation of Rome was viewed with disfavour by public opinion. It was added that the siege operations and the closer investment of the city by the French army, regarded as contrary to the spirit of the suspension of arms, had contributed not a little to the decision of the Assembly. The note wound up by announcing that a counter proposition, which would, it was hoped, facilitate an understanding, would be submitted to me the following day.

As no such counter proposition had reached me on the 22nd, I felt it incumbent, after consultation with General Oudinot, to intimate that we considered that we had exhausted all the means of conciliation, and we accompanied this note by announcing the rupture of negotiations, and stating that we would notify the resumption of hostilities a week in advance. I had calculated with General Oudinot that this interval would enable me to receive from Paris a telegraphic reply at all events to my first despatches.

But the Executive replied to me on the same day that the reason why the promised counter proposition had not been officially transmitted was that fresh bases of negotiation had within the last two days been the subject of verbal communications between the President of the National Assembly, General Oudinot, and the United States Minister.   I inquired as to the truth of this of General Oudinot, and he sent me the following reply :—

"The United States Ambassador (Mr. Cass, son of the general), came to my head-quarters yesterday and expressed his anxiety to assist, unofficially, in making the Roman Government see the necessity of averting the calamities which are impending over the population."

General Oudinot, in expressing his regret that the step which had been taken by Mr. Cass, the son of the American Ambassador, a step to which he had not attached any importance, should have been seized as an excuse for the delay of the Triumvirs to reply to my ultimatum, said he remembered the American agent leaving a paper with him which he had scarcely looked at.   This paper was handed to me, and it embodied the following proposals :—

"Article I.—The Roman Republic, accepting the deliberations of the French Assembly which authorise the despatch of troops to Italy to prevent foreign intervention, will be grateful for the support which it may receive.

"Article II.—The Roman people are entitled to decide for themselves upon their form of government, and the French Republic, which has never questioned this right, will be pleased to recognise it formally, when the constitution, as created by the National Assembly, shall have been sanctioned by a general vote.

"Article III.—Rome will welcome the French soldiers as brethren, but they will not enter the city, unless the Roman Government, threatened by an immediate danger, shall call upon them to do so. The civil and military authorities of the Roman Republic shall fulfil their respective functions. The French Republic guarantees more especially the right which it recognised the Constituent Assembly as possessing to complete and put into working the constitution of the Republic."

This scheme was drafted in the handwriting of M. Charles Bonaparte, the Vice-President of the Roman Assembly, who subsequently gave me a second copy of it. In that which had been handed to General Oudinot by Mr. Cass, the latter had introduced a fourth clause, in which it was proposed that he should sign the agreement as Minister of the United States. It will be readily understood that in view of my instructions I abstained from discussing proposals at nearly every line of which the Roman Republic, which I had not been charged to recognise, was specifically mentioned. I declined to allow them even to be the subject of any written communication on my part.

This incident led me to suspect that the Roman executive power, finding that I was resolved to follow closely the line which I had adopted from the first, was endeavouring to act, irrespectively of me, upon the mind of the general in command, and I was aware that, upon the other hand, a party which had little confidence in the friendly intentions of France, and was disposed to repel all attempts at conciliation, endeavoured to make me appear as an obstacle and a disturbing element. It was openly stated in the clubs that I was another Rossi. The irritation produced by the ringleaders upon a few fanatics led to a disgraceful scene which disturbed a meeting of Frenchmen at the French Embassy. Three men, wearing the uniform of the Roman National Guard, furious at not having met me there and at having " missed their chance," as they phrased it, insulted M. de la Tour d'Auvergne, who had attended in my place.

The complaint which I lodged with the Roman authorities was at first received very coldly, though it is only fair to add that they were not aware of the criminal intentions which, as I had been privately informed, these men had cherished. It was not till the next day, after the discovery of a fresh plot, and after hearing the details spontaneously communicated to them by Colonel Lavelaine de Maubeuge, while I was in conference at head-quarters, that they proceeded to arrest one of the culprits, a Frenchman named Colin. This man was still confined in a dungeon of St. Angelo when I left Rome on the 1st of June.

The general in command kept on writing to urge me to have done with the matter off hand, and though we had both quite agreed that it was indispensable, in the absence of any instructions posterior to my leaving Paris, to gain time and await, at all events for a week, replies to the letters which I had forwarded through M. Forbin-Janson, he sent me message upon message, saying that the generals were pressing him to act; that he had full confidence in me, but that no one shared my sanguine views, and regarded them as illusory. He added that, in the opinion of General Vaillant, the *statu quo* was derogatory to the dignity and interests of the army.

I replied to him on the 23rd :—

" I have communicated to you, before sending them, all the despatches which I have addressed to the Government since my arrival at Rome, and I have to-day sent to Paris, by M. de la Tour d'Auvergne, a general report which I discussed with you yesterday, and to which you raised no objections.

Nor did General Vaillant raise any when he came
to confer with me on your behalf, and I really cannot understand so sudden and complete a change of
front as your successive letters intimate. My conduct has hitherto been most consistent, and as I am
to-day sending to Paris the report which, as was
agreed with you, reserves all initiative on the part
of the Government until its reply arrives, it is
impossible for me to alter my course without the
most urgent cause. At the same time, as my mission
cannot be of any effect if I am harassed in all directions, I am quite prepared to inform the Roman
authorities that I shall retire to head-quarters if
within a week from the present time a solution is not
offered us, either by the acceptance of our original proposals or by a counter scheme which would change
the form of it without altering its spirit. As to illu-
sions, I have none. Nothing surprises me from any-
body, and I am quite equal to meeting all the officious
insinuations designed to make me deviate from the
course which I have adopted.

" The honour of the army is as dear to me, General,
as it is to you; but at the same time I set great store
by the written and verbal instructions of my Government, and of public opinion in France. Do you
desire, yes or no, to enter Rome by force and assume
the offensive without having been attacked, or having
received any formal orders ? When you have once
reached the gates of Rome, and destroyed its walls

with your guns, how are you going to occupy the city?

"Are we at once to give notice to the French families residing in Rome that they had better with-draw if they dread the consequences of an early rupture? And are you prepared to make yourself responsible for the consequences of forcing your pro-tection upon an unwilling population?

"F. DE LESSEPS."

The above considerations and the intentions of the staff, as manifested by General Oudinot's corre-spondence, made it necessary for me to take up my residence for a time at head-quarters. I went there on the 24th, and the General gave me a room in the Villa Santucci, where he was living himself, and where he received me very kindly. From there I lost no time in addressing, after reading it to him, to the Constituent Assembly a message explaining our draft of agreement, the clauses of which we retained, with this addition :—

"The French Republic guarantees against all foreign invasion the Roman States occupied by our troops."

This clause was the reproduction of an order of the day, addressed by the general in command to the commanders of the different corps when the march of the Austrians into the States of Rome was made known, the which order he had authorised me to copy from his correspondence. He informed me, however, that he had kept it secret up to the present, and that if I thought it expedient to make use of the expression in our draft of agreement with the Romans, he saw no objection.

General Oudinot again promised me to await the final decision of the French Government, but he was still much worried by the complaints which the generals under him made as to the inaction of the army. It was agreed that I should explain, at the next council of war, the political situation in which we were placed. The generals who were present were Vaillant, Rostolan, Regnault de St. Jean d'Angely, generals of division; Gueswiller, Le Vaillant, and Mollière, generals of brigade, and two others, whose names I have forgotten; in addition to Colonel de Tirion, the chief of the staff.

I read part of my despatches to the Ministry as well as the accompanying documents, and I added that I had resolved to formally oppose a resumption of hostilities against Rome pending the arrival of orders from the Government, basing my declaration upon the latest instructions sent to the commander-in-chief by telegraph on the 10th of May. Several generals asserted that a simple demonstration would suffice to ensure the opening of the gates, that at the most there would be no need to do more than knock down the corner of a single wall, and that there would be no real resistance. I maintained that they were mistaken; that once hostilities had been begun we should be led on to shed a great deal of blood and destroy many buildings; that the resistance would be an obstinate one; that we should be obliged to lay a regular siege; and that though we should un-doubtedly end by accomplishing our aim, nothing being impossible to a French army, I would never take upon myself the responsibility of evils easy to foresee; and that the general in command had re-ceived no instructions which authorised him to assume himself the responsibility contrary to my expressed opinions.

General Oudinot submitted to the consideration of the Council the following question:—

" Is it desirable to abandon the negotiations and resume the attack upon Rome, without regard to the conclusions of M. de Lesseps and without awaiting fresh instructions ? "

The majority was at first inclined to vote for attacking at once; but General Mollière, well known for his exploits in Algeria, the youngest member of the Council, had not yet spoken. Being asked to give his opinion, he said that he was sorry, as a soldier, not to be able to vote in favour of immediate action, but that he felt it difficult to dissent from my opinion as to the expediency of waiting for fresh instructions. It was accordingly decided to maintain the *statu quo*.

During my stay at head-quarters I kept up, through my private secretary, M. Leduc, daily intercourse with the Roman authorities and several influential persons not in office.

On the 24th of May I addressed a letter to the French residents in Rome, informing them that during my absence the French flag would continue to float over my hotel and the French public buildings, and that they could display it from the windows of their own houses if they pleased. I added that they were to apply to M. Gerando, the Chancellor of the Embassy, in case of need; and impressed upon them the importance of their being reserved and discreet in their intercourse with the Romans.

I also forwarded to the Minister of Foreign Affairs in Paris a letter which I had received from the Triumvirate on the 26th, urging that the French troops, while remaining in occupation of Civita Vecchia, should not enter Rome, or interfere to prevent the Romans from taking the field against the Austrian and Neapolitan forces. To this I replied by the following letter, a copy of which I forwarded at the same time to M. Drouyn de Lhuys:—

" FRENCH HEAD-QUARTERS,
" VILLA SANTUCCI, <i>May</i> 26.

" Gentlemen,—I have received with much satisfaction the letter which you did me the honour to address me yesterday. The explanations which I have already

given to the three commissioners of the Roman Constituent Assembly, and the communications which I
have thought it incumbent upon me to make to the
Assembly itself, meet, without exception, all the objections raised in your note, and whenever you see fit to
complete the negotiations by sending your commissioners invested with the necessary authorisations,
it will be very easy, in my opinion, for us to come to
a complete understanding and settle the basis of a
definite arrangement, which must, of course, be one
such as will quite satisfy the two contracting parties.
This declaration, which my private secretary will be
able to supplement with a few verbal observations,
will, I am sure, dissipate the unfortunate misunder-
standings which may have hitherto arisen upon either
side. For my own part, I have been, still am, and
shall still remain desirous of clearing up the obscurities
which have hung about the question, just as I hope
that my language will destroy any lingering doubts
which you may have felt as to the result you have in
view. There is only one point which could in any way
make you feel anxious, and that is the idea that we
are intent upon imposing on you by force the obligation
to receive us as friends. Friendship and violence do
not go together, so that it would be illogical of us to
begin by cannonading you as a preliminary to getting
you to look upon us as your natural protectors. Such
a contradiction in terms does not enter into my intentions, or into those of my Government or of our army.

This was the purport of what General Oudinot said yesterday in my presence to the Roman deputation which came to offer, in your name, a present of fifty thousand cigars and two hundred pounds of tobacco for our soldiers, and his remarks must have removed any doubts which may have lingered in some minds. But so long as we begin to understand each other, there is no good in going back upon the past. Let us rather concern ourselves with the present and the future. You will, I repeat, find us entirely disposed, by our words and written statements, to give you the explanations and guarantees which your natural susceptibility from a national point of view may demand. It would ill become Frenchmen, noted for their unlimited devotion to their country, to blame other nations for defending their territory against their real enemies, or to compel you to do the very contrary of what they would always do in their own case.

"Accept, &c.,

"F. DE LESSEPS."

CONFERENCE HELD AT THE HEAD-QUARTERS, VILLA SANTUCCI, BETWEEN M. DE LESSEPS AND M. DE RAYNEVAL.

Note from M. de Rayneval to M. de Lesseps.	*Note from M. de Lesseps in reply to M. de Rayneval.*
May 27.	
1. You have had the kindness to take me into full confidence as to your ideas, your	1. M. de Rayneval, starting from a different standpoint from mine as to the attitude to

projects, and your action. In return for the confidence which you have shown me, I feel that I cannot do less than treat you with equal frankness.

2. My private views are of little consequence, but the Government of the Republic, in requesting you explicitly to concert your action with the plenipotentiaries of the Gaëta Conference, was evidently anxious to avoid speaking with two voices. I am obliged to say that you have not been influenced by this consideration, a very important one in my eyes, as it involves the honour and good faith of the country.

be taken in this business, should be consistent with his principles, as I have been with mine. His reservations do honour to the perspicacity of his eminently politic mind, and to the hereditary loyalty of his disposition.

2. My instructions were to the effect that I was to concert with MM. d'Harcourt and Rayneval upon all matters not requiring an immediate solution. I have communicated with M. d'Harcourt whenever he has come to head-quarters, where I have made a point of meeting him, despite my incessant occupations, to communicate to him not only all that I had done, but the ideas which inspired me in my action. At the same time, I sent to M. de Rayneval at Gaëta duplicates of my earlier despatches to Paris, and I should have gone on doing so had he not, much to my satisfaction, come to head-quarters. I have kept nothing back from him; I have let him know all the powerful motives, public or secret, which have directed my conduct, and he must have carried away with him the conviction that if we were divided as to our views, I was not less anxious than he to maintain intact the honour and good name of our country. As to consulting

MM. d'Harcourt and Rayne-
val previous to each of my
steps, which, day by day, hour
by hour, minute by minute,
demanded immediate and vary-
ing decisions, it would have
been impossible. Even General
Oudinot, who was more directly
concerned with me, did not
expect this.

3. I also have to point out
that you act not only without
any regard to antecedents, but
solely guided by your inspirations and without any written
instructions from Government.

3. The antecedents which at
the outset guided the course of
M. de Rayneval should un-
questionably have been taken
into consideration by me, but
they could not serve as an invariable rule for me, because,
upon the one hand, the Holy
Father, who will certainly recognise the necessity of putting
himself in our hands, has never
done as we advised him; because he has followed advice
diametrically opposed to ours;
because his court has become
a modern Coblentz, from which
French influence has been ex-
cluded; and because, upon the
other hand, the facts which
had marked the début of the
French expedition, after the
occupation of Civita Vecchia,
had complicated the question
and produced a quite different
situation. This situation, the
exact nature of which the
Government could not understand when I left Paris on the
8th of May, prevented their
giving me precise and detailed

instructions. I was merely
told to " devote my attention
exclusively to the establishment of relations with the
Roman authorities and inhabitants. . . . You will be guided
in your judgment by circumstances."

4. Being quite out of harmony with d'Harcourt and
myself, deriving all your force
and authority from the circumstance—the importance of
which I quite admit—that you
have more recent information
as to the intentions of our
Government, you are quite
master of the situation and
paralyse the army.

4. I have reported to the
Government all that I have
done to check the eager-
ness of the army—an eager-
ness which is greatly to its
credit, but which, if not mode-
rated in present circumstances,
would have led to an irrepar-
able catastrophe, the result of
which would have been, in my
opinion, to destroy our influence and aid the schemes of
our enemies at home and
abroad.

5. You have gone so far all
at once that the risk of putting obstacles in your way is as
great as that which you have
voluntarily incurred. Moreover, you have appealed to the
supreme judgment of the
Government, so that it is but
right to await their decision,
which will, I trust, be not long
delayed.

5. My object was to ascertain the truth, to let the
Government know it, to clear
the way for them, and to enable
them to come to a final decision,
the announcement of which I
am awaiting not less eagerly
than M. de Rayneval.

6. It is possible that the
Romans will open the gates of
the city to us. They will be
the less likely to do so when
they see that the army is not
preparing to act. But owing
to the conditions which you

6. I had no wish that the
Romans should at once open
their gates, as it was better to
give time for the passions which
had been excited by the events
of April 30th to subside. All
the conditions which I had at

have laid down, the solution of the question, it seems to me, is more likely to be retarded than advanced.

7. I protest to the utmost of my convictions against your proposals. They involve not only the recognition of a Government which the French Republic has expressly declared that it will not recognise, but an offensive and defensive alliance with this Government.

This is the first and a very grave infraction upon what I believe to be the instructions of our own Government.

8. In reality we are throwing down the glove, not only to the three Powers which have declared war against the Government of the Republic,

first proposed could not be accepted, and I should have been sorry if they had been; among others, that which consisted in having the military service of the city performed jointly by our troops and those of Rome.

The general in command was very anxious that this should be done, but I have always declared that I regarded this as a danger, because it would involve us in questions of Roman administration to a greater extent than I deemed advisable, and because it would entail our taking upon us a part of the inheritance of the present executive.

7. In my proposals there is not a word as to the recognition of the Roman Republic, and that is so true that they are regarded by Mazzini himself as inacceptable, and as containing in reality nothing more than the substance of General Oudinot's first proclamation before Civita Vecchia. I was instructed to negotiate with the people and authorities of Rome, and in doing so I have merely conformed to the written instructions of our Government.

8. By our conduct here we are in no way throwing down the gauntlet to the three Powers which have declared war against the Government

and which have the support of all Europe, but to a Power superior to all these, and destined to play a very important part in our internal destinies: I mean the Papacy.

This is a second and not less serious infraction of the rules laid down by the Government of the Republic, which has not declared war upon Austria, which is solely desirous of placing itself in such a position as to exercise the due influence of the French Republic in the ulterior regulation of the affairs of Rome.

of Rome. Naples has seen fit to take military action, but we never promised to side with her, and when General Oudinot was called upon by M. de Ludolf to come to a decision, he left him with no doubt upon the subject. A telegraphic despatch of May 10th, sent by the Minister of Foreign Affairs to General Oudinot, instructed him to inform the Romans that we did not intend to join with the Neapolitans against them. We do not therefore throw down the glove to Naples, any more than we do to Spain, whom I was myself instructed to inform that her ambassador at Gaëta had very unwisely separated her cause from ours, and that in taking part, together with Austria, against us, he was serving the interests neither of the Papacy nor of Spain. With regard to Austria, her principles are so different from ours that it is very difficult for us to be agreed, while a pretence of agreement would not be of any use to you, and would alienate for ever the Roman populations. A decision must therefore be come to, and if we are to avoid war with her it is by going on as we have begun, unless new and unforeseen circumstances should arise, and by fortifying day by day our

military political position in
the Roman States. . . .

9. In uniting with the enemies of the Pope, you inevi-
tably drive him more than ever
beneath the exclusive influence
of Austria. This assuredly is
not a result which it is desirable to attain.

9. Far from uniting with the
enemies of the Pope, we prove
to him, on the contrary, that
we are the only nation sympathetic to the Roman populations which conceive their interests in a just and liberal
measure ; and if at this moment his spiritual influence
even is compromised at Rome
by his imprudent friends quite
as much as by the hatred of
his enemies, he will be convinced one day that we alone
can open for him the doors of
St. Peter's, and lead him along
a way strewn with flowers.
He will understand that access
to that basilica would be closed
to him if he had to reach it
along a path sprinkled with
the blood which he had caused
to be shed. Some sincere
friends of His Holiness have
encouraged me in the path
which I have followed, and
they have strongly urged him
not to raise any difficulties in
my path.

10. Is it really the desire of
France to offer her hand to a
Government which began its
career by murder . . . .
and which relies for its existence upon our internal
troubles ?

And you must bear in mind
that directly we recognise this

10. It is no more accurate to
say that the Roman Republic
is responsible for the murder
of M. Rossi than to make our
Republic of '48 responsible for
the crimes of '93. The Roman
Republic—which, moreover, I
have not been charged with
recognising—has succeeded, by

Government, we cut from under our feet the only ground upon which we can maintain ourselves. If, in our eyes, this Government exists, if it is the outcome of the free choice of the nation, we are bound to support it. It would not be right for us to do what we can to bring about its fall except while it remains what it now is, the work of a faction composed chiefly of foreigners.

11. You paralyse the army, forgetting the maxim, "Si vis pacem," &c. You expose it to demoralisation and to disease. The army, which is anxious to show what it can do and to shed a fresh lustre upon the French name, is condemned to capitulate . . . .

an almost unanimous vote, to the Government which had been the direct heir of the murder of Rossi; and it was proclaimed by an Assembly whose mission it was to choose the form of government which it preferred. This is a fact, the consequences of which I am not called upon to discuss.

11. I do not paralyse the army, but do all I can to prevent its admirable ardour causing it to deviate from the right path. The army will have deserved the gratitude of the country by reserving this ardour to combat the enemies of our independence and influence, instead of committing the deplorable error of employing it to make breaches in crumbling walls and to destroy the finest monuments of ancient and modern genius. My despatch No. 6 indicates how our army, so brave, so well disciplined, and so well commanded, may maintain its position, and fortify and improve it by a change of quarters, in the event of Rome not opening her gates before the season of fevers sets in. This project ought, by good rights, to be carried out from the very day that an arrangement is come

to between the Romans and
the French. To avoid any
unfortunate contact between
the two forces, and to place us
in such a position that we
could retire without inconveni-
ence whenever France might
require her troops, a strong
position in the city, where the
head-quarters could be established with the necessary forces,
will be the object of special
stipulations in the event of an
agreement being come to. I
have pointed out to M. de
Rayneval, upon a map of Rome,
the advantage we should derive from occupying, upon
Monte-Pincio, a part of the
Académie de France and all the
buildings attached to the splendid convent of Notre-Dame-
du-Mont. These French pro-
perties form a very good body
of military positions. The steps
of the convent descend into the
interior of the city, and up to
1815 any man pursued by the
police had only to put his foot
upon the first step of these
stairs to enjoy the privileges of
the inviolability of French territory.   The sisters of the
Sacred Heart, who now inhabit
it, are only tenants of the
French Government, and they
have two other very fine pro-
perties in Rome, to which they
could meet.

12. While our army remains

12. The principal canton-

inactive under the walls of Rome the Austrians advance, and the Pope might very possibly go and re-establish at Bologna, under their ægis, the seat of authority.

Beneath the walls of Rome, and even if we were to agree to share the few posts in the city which the authorities might condescend to offer us, should we be in a position to hold, either to the Austrians or to the Pope, the language which it is fitting that France should employ? Our only resource in dealing with the former would be force, which would be utterly useless in regard to the latter.

ments of our army being established at Frascati, Albano, and in the neighbourhood known as the Camp of Hannibal, we should maintain our free communications with Civita Vecchia, the routes to Florence, Bologna, and Fiumicino by the Tiber ferry, or the post which General Oudinot has already prepared, and we should have a new and better communication with the sea by the Porto d'Anzio (Portus' Neronis). In such a situation, which is being thought out at the present time by General Vaillant, we could not be regarded as having wasted our time or remained inactive. The march of the Austrians need not cause us any uneasiness. As to the fear of seeing the Pope establish his see under the ægis of Austria at Bologna, an open and defenceless city, I do not think that there is any foundation for it.

13. Primary assemblies in countries such as this have not the moral force which they generally have with us, because every one knows that in Italy the populations are incapable of expressing their wishes in this way. In leaving them to decide the future fate of the Roman States, we implicitly declare that we no longer concern ourselves with the sovereignty of the Pope,

13. In declaring to the people of Rome that we do not dispute their right to choose freely their form of government, we do not indicate how this free choice is to be used; and if we do not just at this moment raise the questions relating to the interests of the Holy Father, it is because we should regard it as very imprudent to do so prematurely, being convinced that time alone can

whereas we have solemnly declared in the face of Europe that we would respect the territorial divisions recognised by the treaties.

create a spontaneous movement in his favour. As to bringing him back by force, no one can deny that his restoration would not be durable unless it was maintained by the same violent means which had effected it.

14. I am not in the least alarmed by the efforts of the Protestant missionaries. They may create a scandal, but that is all.

14. I saw a good deal of what the Protestants are doing in Rome ; the danger is a real one. It might perhaps be only a passing one, fated to diminish or disappear altogether when the help upon which they now rely is no longer to be had. But as, after all, we have to do with things as they are at present, not as they have been or may be again, it is necessary to combat and to beat down the hostile elements opposed to us.

15. One word again as to the kingdom of Naples. You expose it to the invasion of armed bands whom our inaction leaves at liberty. Is the French Government desirous that the agitation in Italy, no sooner suppressed in the north, the centre, and Sicily, should break out anew in Naples?

15. It is not our fault if the kingdom of Naples is exposed to being invaded by Garibaldi's forces ; but it is owing to the imprudence of the Neapolitan troops in advancing into the States of Rome.

16. I have said enough to show how sad at heart I feel in leaving Rome with matters in such a pass as they are. I should deplore as much as you that the way of the Papacy should be one of bloodshed and ruin. Such must not be

16. In the situation as it now is, an attack upon Rome would, as I hope that I have demonstrated in my correspondence with the Ministry, have led to great disasters and would have been aimless.

the case. In my opinion, a very firm attitude, an attack which, with no worse results than the destruction of a few tottering walls, would have made us masters of the Roman head-quarters, would have led the population to decide in our favour. . . . We should at all events have been in a strong, healthy, and satisfactory position, both as regards our national pride and the political necessities of the day. Sooner or later we should have been received unconditionally, and been allowed to dictate our own terms. We should not have had to struggle, as we now shall have to do, if you succeed, against engagements which we cannot fulfil, and which will most seriously compromise us in the eyes of all Europe.

17. I deem it my duty to decline formally and fully all responsibility for what has taken place since your arrival. But I will not conclude without rendering homage to your zeal and good intentions, without begging you to see in my outspoken frankness nothing save a proof of my confidence in your character and of my now long-standing affection.

DE RAYNEVAL.

17. I willingly assume full responsibility for what is being done before Rome. I do not ask anyone to share it with me, and I honour M. de Rayneval for arriving at convictions different from mine as the outcome of ideas which spring from the same patriotic source as my own. I thank him for his frankness, which harmonizes so well with my own, and with the sincere affection which I feel for him.

F. DE LESSEPS.

Different as were our views as to the pending negotiations and the hopes which they might evoke, M.
de Rayneval, indicated (see clause 5) the essential
point. " You have appealed to the supreme judgment of the French Government, so that it is but
right to await their decision."

Upon the 28th General Oudinot held, a few miles
from head-quarters, a review of 10,000 men, most of
them belonging to corps recently arrived from France,
and he asked me to accompany him. On the 29th
I arranged with him to address a very pressing communication to the Roman authorities.

The announcement of the forward march of the
Austrians, the desire to give satisfaction to the army,
and the hope of seeing an honourable compromise
agreed to by the majority of the Assembly, which, I
was assured, was very favourably disposed, induced us
to send to Rome what was practically an ultimatum to
the Assembly, the Municipality, and the Triumvirate
each of them receiving a copy from M. Leduc, my
private secretary. This declaration was as follows :—

" The undersigned, F. de Lesseps, Envoy Extraordinary and Minister Plenipotentiary of the French
Republic ; Considering that the march of the Austrian
army upon the Roman States changes the respective
position of the French army and the Roman troops ;
Considering that the Austrians, in advancing upon
Rome, might seize positions which would be threatening for the security of the French army ; Consider-

ing that the prolongation of the *statu quo*, to which General Oudinot, the commander-in-chief, had at his request consented, might become injurious to the French army ; Considering that no communication has been addressed to him since his last note to the Triumvirate, dated the 26th inst. ; calls upon the Roman authorities and Constituent Assembly to give a definite answer to the four following articles :—

" Art. 1. The Romans request the protection of the French Republic.

" Art. 2. France does not dispute the right of the population to pronounce freely as to their form of government.

" Art. 3. The French army will be welcomed by the Romans as a friendly one. It will occupy the cantonments which it may deem the most suitable, both for the defence of the city and for sanitary reasons. It will not interfere at all in the general administration of the country.

" Art. 4. The French Republic guarantees the territory occupied by its troops against any foreign invasion.

" In consequence, the undersigned declares, in agreement with General Oudinot, that if the above articles are not at once accepted, he shall regard his mission as terminated, and the French army will resume its liberty of action.

" F. DE LESSEPS.

" (Countersigned) OUDINOT DE REGGIO.

" HEAD-QUARTERS, VILLE SANTUCCI,
" *May* 29, 1849."

Subsequently it was agreed that a further delay of twenty-four hours, expiring at midnight on the 30th, should be granted.

After the departure of M. Leduc, in the night of the 29th to the 30th, I remarked that there was a great stir among the staff of the general in command, and that preparations were being made for some movement the next day. I at once handed to the General, in whose house I still resided, a note in which I said, "In the event of your deeming it your duty to seize, by surprise or otherwise, positions inside Rome, or even just outside its walls, without previously consulting me, I think it only right to disclaim all responsibilities for the political consequences which may result from it. Until orders arrive from the Government, either blaming or approving of my conduct, it is not in keeping with my mission that you should alone determine upon any measures, military or otherwise, which might compromise our Government or implicate our country in a cause which I deem to be ill-advised and dangerous."

It should be added that I had not, up to this date, heard a word from the Minister of Foreign Affairs, nor had General Oudinot. The embarrassment which this silence might cause had been pointed out by me in all my despatches, notably in those sent through M. de la Tour d'Auvergne and one or two other messengers.

I knew that the General had sent for several

generals commanding army corps, to give them his instructions with regard to the attack on Rome which was to begin at midnight. Having learnt that the General had determined to take no account of my advice, I thought it my duty to submit to General Vaillant, who had recently arrived from Paris, whose coming had been telegraphed to me by the President, and who might therefore be supposed to be in possession of the latest views of the Government, my serious motives for opposing the proposed attack, and for believing that the immediate occupation of Rome was pregnant with danger. I wrote to him as follows:—

"*May* 30, 1849.

"I was hoping to have come to see you this morning, in order to communicate to you confidentially the result of my latest conference with M. de Rayneval. But I have not been able to find the time, being busy preparing a note for the purpose of showing that, from the political point of view, the necessity of despatching a division which would be distributed over Albano, Frascati, and Marino, on account of the recent landing of 4,000 Spaniards at Gaëta. It is said that they will give fresh courage to the Neapolitans and that the campaign will be resumed. It is our duty to anticipate them in the occupying of the encampments around Rome which they might otherwise seize, and, by thus anticipating them, make our-

selves the sole masters of the city.   In this way we
do not compromise our Government by a premature
entry into Rome, by remaining in a city which even
its inhabitants abandon in summer. . . . . We shall
be the real masters of Rome if we surround it with
troops, and the Government of the Republic, *which does
not desire our entry into Rome except by agreement with
the inhabitants*, will thank you for having contributed,
by the wisdom of your counsels, to bring about the
triumph of the true and only policy stripped from all
petty questions of *amour propre* and vainglory."

At three o'clock the same day (that is, nine hours
before the expiration of the delay agreed upon) I received the replies of the President of the Roman
Assembly and the members of the municipality to
terminate the negotiations, to prevent France assuming towards Rome the part of Austria, and to put an
end to the misfortunes with which a peaceful city, the
home of so many monuments and of the arts, was
threatened.   At the same time the Triumvirate sent me
a note and a set of counter proposals.   It was evident
that the General-in-chief and myself could take the
counter project of the Roman authorities into consideration, discuss it, and only break off all negotiations,
pending instructions from Paris, in the event of our
finding it impossible to come to an understanding.
The simple dictates of humanity would have led us to
do so, even if this course had not been indicated by

the situation in which we found ourselves placed, and which had not varied since the 15th of the month, the day of my arrival; that is to say, our Government, which had long since been called upon to decide, had not sent us a word either by courier or by telegraph. Thus General Oudinot was still bound by the despatch of the 10th, which did not allow him to attack. I put together the letters of the Roman municipality, the Assembly, the Triumvirate, as well as the project which I had annotated. I added to it the following memorandum, and requested Commander Espivent to hand the whole to General Oudinot:—

" MEMORANDUM.

"Having started from Paris while the impression produced by the events of April 30th was still fresh, and having come for the purpose of treating with the Roman populations, I have no need to recall the fact that I have never allowed for an instant my cause to be treated as a distinct one from that of the worthy Commander-in-chief.

"I was not blind to the difficulties which I should experience in bringing people to believe that the intentions of the French Government and of its general were the same after as they were before the 30th of April. But I have at last succeeded. I am disposed to sign at once, with a few modifications and the rejection of Art. 3 relating to the recognition of the Roman Republic, the counter proposal sent me by the

Triumvirate, and approved by the Roman Constituent
Assembly, as well as by the senators and conservators
of the Roman municipality, in the belief that this act
is calculated permanently to strengthen French influence in Italy, and maintain the unspotted honour
of our army and of our glorious flag.

"*May* 30, 1849."

M. Espivent shortly afterwards returned me, by
General Oudinot's order, the bundle of letters, which
he had not had time to read, being very occupied with
the details of his service and the orders he had to give
to the army. He had added, however, that I could
explain my views presently at a council meeting of
the generals which took place at four o'clock. Despite
my reading of the above documents and of my observations as to the absence of orders from our Government and of the perilous nature of the situation,
nothing could alter General Oudinot's resolve, while
the manner in which he expressed his views made all
discussion impossible, and compelled me to advance
my official position in order to check the outbursts of
temper which I plainly said that I would not stand.

A similar scene occurred again at night in the
presence of one of my friends, Captain Chaigneau, in
command of the frigate *Magellan*. But my firmness,
coupled with the dropping of my hand on to the hilt
of my sword when the General talked of having me
placed under arrest, fortunately had the effect of re-

voking at the last moment all along the line of
advance posts the order which had been given for
an immediate attack.

However, as I was afraid that these orders would
not arrive in time to prevent deplorable consequences,
I made it known in Rome, whither I at once proceeded, that there was no occasion to feel uneasiness
at our movements, which were only intended to
enable us to make sure of the positions which foreign
armies marching upon Rome might seize. But for
this advice we should not have been allowed to
occupy Monte Mario without resistance. The aide-
de-camp sent to countermand the occupation of it
arrived too late. I returned during the night to
head-quarters.

Early the next morning General Oudinot, having
heard that I was making my preparations to return to
Rome, sent one of his aide-de-camps to ask me to come
and see him before I started, and I replied, that as I
had a final note to hand him, it was my intention to
have done so. The General told me that he much
regretted what had occurred ; but I would not let
him finish his sentence, and grasped the hand which
he had put out. I told him that I was going into the
city in order to complete the negotiations upon the
basis of our annotated project. In order to show him
that I had foreseen the possibility of my not stipu-
lating for the immediate occupation of Rome by the
French army, I took the precaution of reading him

the following notes, the original of which I left in his hands :—

"It is expedient, until their passions are calmed down, to avoid all contact between the army and the people, whom so many causes maintain in a state of effervescence.

"The air of Rome is insalubrious in summer.

"The sojourn in Rome of any part of our army might, as I have not ceased to declare since my first conference with M. d'Harcourt, involve us more deeply in internal Roman questions than we should wish to compromise our Government. M. de Ray-neval was of opinion that the military service conducted jointly by our troops and those of Rome would be most prejudicial to our interests.

"The general embarrassment arising from the empty state of the treasury, the accounts which have to be rendered for the immense sums of money spent without any sort of check, are so many insurmountable difficulties in the way of the present Government should it become consolidated, or for any other power which might succeed it. The administration, being bound to offer some sort of justification to the people for all the financial embarrassments which will have arisen, will no doubt seek to attribute the effects of it to the French occupation. This forced occupation might, moreover, have the result of keeping up a feeling of irritation against, and a desire of vengeance upon,

our soldiers, which the slightest incident might bring
to a painful head. Is it not better to wait till these
feelings have calmed down? Then the people themselves will come to ask the French soldiers to enter
the city, and perhaps in a few days. It is all very
well for us to declare that we shall not interfere in the
general administration of the city; but a military
occupation would be certain to entail this; and how
can we tell that, once entering Rome by force, we
should not be obliged to employ force to maintain
ourselves there, or that we should be free to withdraw
our troops just when it was convenient or necessary
to do so? It seems unnecessary that I should dwell
further upon this situation or upon the other grave
matters to which I have made allusion.

"F. DE LESSEPS.

"P.S.—M. de Rayneval writes to me from Gaëta,
under date of the 28th, that, according to letters from
Rome, the city has resolved to defend itself if attacked.
He adds, 'The moderate party would not care to face
the perils of a reaction of which they hope that we
shall spare them the cost.' This was just the view I
had taken the day I arrived in Rome. I am very
glad that this view, so opposed to all that had been
said before, is confirmed by the news from Gaëta."

The General sent me a note, written in pencil, in
which the aide-de-camp sent to communicate with the
commander of the column ordered to take up a position

on Monte Mario said that he had not arrived in time to countermand the movement. This note was intended to reassure the Roman authorities as to our intentions, and it enabled me to explain the occupation of Monte Mario to the Triumvirs.

Upon my arrival in Rome, I found them very much concerned about the occupation of this important point, and numerous complaints had been addressed to them from all parts of the city. I gave them to understand that it was all a mistake, and they undertook to reassure the population.

I handed them a final proposal, which I had thought out with great care, and which I believed to be as faithful an expression as possible of my instructions and of the intentions of the Government, as conveyed by the speeches of ministers at the sittings of April 16 and May 7 and 9. My proposals did not respond to the hopes of the Triumvirs, because they made no mention of the Roman Republic, which I could not recognise, and because they only guaranteed from foreign invasion the territories occupied by our troops. They were, however, accepted as a necessity, and Mazzini told me that if the Assembly agreed to them it would be a proof of the great confidence which I had inspired as to the intentions of the Government; for if these intentions were not such as I described them, it would be very dangerous for the Romans to agree to my project. "For," he added, "the positions of which we are about to facilitate your occupation of, and the privilege which you reserve for yourselves of only repulsing our foreign enemies if they come directly in contact with yourselves, leaves our political existence at the mercy of your good faith."

The Roman Assembly, convoked the same day, in a sitting with closed doors, adopted my scheme all but unanimously. This result was announced to me by the Triumvirs in the following letter:—

"The result of the long discussion in the Chamber is as follows:

"Art. 1. The support of France is assured to the inhabitants of the Roman States. They regard the French army as a friendly one, come to aid in the defence of their territory.

"Art. 2. By arrangement with the Roman Government, and without in any way interfering in the administration of the country, the French army will occupy the outward cantonments most suitable for the purposes of defence and for the good health of the troops. Communications shall be free.

"Art. 3. The French Republic guarantees against all foreign invasion the territories occupied by its troops.

"Art. 4. It is understood that the present arrangement is to be submitted for ratification by the French Republic.

"Art. 5. In no case shall the effects of the present undertaking lapse until a fortnight after official communication of the note having been ratified.

"There appear only to be very slight changes in the wording of this project which can give rise to any objections. If this is so, all that remains to do is to settle the means and the form of the communication.

"It is impossible to select this evening a deputation of the Chamber to send to head-quarters, but we could arrange that the senator of Rome, Sturbinetti (president of the municipality), should form part of that which we would despatch to-morrow.

"The basis of the arrangement being agreed upon, we would proceed at once to the election of the plenipotentiaries, who would proceed to the camp to settle all details, such as the choice of cantonments, which would be the first result of the convention, and to invite the general in command and his staff to come and reside at Rome with a guard of honour.

(Signed) "J. MAZZINI, A. SAFFI, C. ARMELLINI."

We did not think it wise to fix in advance the number of men who should form the guard of the General; we merely agreed that, by the clause which established liberty of communications between the city and its outer cantonments, the General would be at liberty to let all the forces of his army, one after another, pass through the city.

I then had three copies of the convention drawn up, which were at once signed by the Triumvirs, with the assent of the Roman Assembly and the municipality. The General, to whom I read the proposed convention, declared, directly I came to the question of cantonments, that he would not sign it. He refused to listen to any explanation, and as he spoke in a tone which I did not at all like, I was obliged to reply to him in a way which cut short the whole discussion. Being convinced that this project met all the necessities of the case quite as well as the previous one which he had approved of, and that in some particulars it was preferable to the other, it was impossible for me to give way, especially as I knew that it was General Oudinot's intention to take advantage of the rupture of negotiations and not wait for official orders from Paris; for I had no doubt whatever as to his being secretly in harmony with those members of our Government who belonged to the reactionary party. I bore in mind the vote of the 7th of May, which had led to my mission; I had just read over again the speech delivered by the President of the Council at the sitting of May 9th, according to which I was to be "the faithful and exact expression of the ideas of the Assembly and of those of the Government, with respect to the aim and character which the French expedition was to maintain to the very end and in spite of all eventualities whatsoever." I was acquainted with all the means of defence in the city, and I was certain that until we had received reinforcements and fresh material of war we could not carry it by assault; that the resistance would be prolonged and general. The further time I had spent in

Rome only confirmed me in the opinion which I had expressed on the 15th, the very day of my arrival. I saw that our Government was being entangled in the most deplorable manner, and I felt that, pending fresh instructions, it was our duty to conform to the telegraphic despatch of May 10th, which authorised our entering Rome, "if we were agreed with the inhabitants," and which permitted us to attack "if we were compelled to do so, and in the event only of having the fairest prospects of success."

Could I tear up my own agreement and dispense myself from submitting it to the consideration of my Government, which, after all, had *exclusively* instructed me to negotiate with the *Roman authorities* and to conclude partial arrangements with them?

In short, I considered that the essential thing to be done was to prevent an immediate and imprudent attack; that my provisional arrangement would at all events have the effect of suspending it for a few days, and of leaving Government the choice between peace and war. I had, moreover, informed the Minister of Foreign Affairs, by telegram on the 22nd, that if it had been resolved to adopt a line of policy differing from that which I considered to be the outcome of the vote of May 7th, I wished to be recalled.

Moved by these considerations, I determined not to be deterred by the opposition of General Oudinot. I signed in his presence and left on his table one of the three copies of the convention. I informed him that

I should send a second one to Paris by Colonel
Lavelaine de Maubeuge, and that the third would
remain in the hands of the Roman authorities.
Returning to Rome the same night, I received the
next morning (June 1) a letter from General
Oudinot, protesting against the arrangement concluded with the city. The Triumvirate subsequently
sent me a copy of a communication by which he
declared that what I had done was null and of no
effect. I replied that so far as concerned myself I
adhered to the arrangement, subject to the ratification
of my Government, and that I was about to start for
Paris.

Subjoined are the documents relating to the final
incidents of my mission.

"ROME, *June* 1, 1849.

"M. le Ministre,—My despatch of the 29th ult.
was accompanied by a declaration, in the form of an
ultimatum addressed by agreement with General
Oudinot, to the Roman authorities. The result which
I hoped for was attained ; nine hours before the expiry
of the time fixed I received from the President of
the Constituent Assembly and the members of the
Municipality, composed of all the most distinguished
men in Rome, very satisfactory replies, the Executive
Power, which had been appointed to arrange with
me, sending me at the same time a counter project.
This document, upon the margin of which I at once

annotated my remarks, was of a character to be taken into consideration, and proved to me that the persons who a few days before were resolved to throw difficulties in the way of all attempts at conciliation were no longer masters of the situation. The accompanying documents and the verbal explanations which you will doubtless receive from Colonel de Maubeuge will show you the unexpected and to me very painful behaviour of General Oudinot, who has hampered my operations and was on the point of foiling them altogether. In spite of an opposition and of difficulties which I ought not to have had to meet, an arrangement has been come to between the Roman authorities and myself.

" I have the honour of sending to you one of the three original documents, of which a second is in the hands of the Triumvirate and a third in those of General Oudinot. You will observe that I have suppressed Article II. of the previous project, in which mention was made of the right of the Roman people to select their own form of government, a right which it would be hard for us to question, but the recognition of which, according to M. de Rayneval, had very considerably increased the susceptibilities felt at Gaëta. . . . . My position at head-quarters would be as false as it would be at Rome. I therefore consider my mission as at an end, or at all events as being of necessity suspended, and I am about to make over to M. Gerando the supervision of the

interests of the French who are still in Rome. . . . .
My preparations for starting will not take long, and
I shall follow close upon the bearer of this letter.

(Signed)   "F. DE LESSEPS."

The Triumvirate to M. de Lesseps.

" ROME, *May* 30.

" Sir,—We have received the declaration of May
29th which you have done us the honour of addressing
us. The Assembly, to whom a copy was also sent,
having confirmed its original decision, which delegated
us full powers to treat, it is for us to reply to you.
We do so with all due speed. If we have not sooner
answered your letter of the 26th, it was because, as it
did not contain any proposal on the part of France,
nor any reference to that which we had made to you,
it did not seem to call for any urgent reply.

" We have carefully examined your declaration,
and we append the modifications which we deem it
our duty to submit to you. As you will at once see,
they relate more to matters of form than of substance.
We might say a good deal in advocacy of the changes
which we propose—changes which, as we believe, are
demanded not only by the nature of the mandate
which we hold from the Assembly, but also by the
clearly expressed wishes of our fellow-citizens, without
whose assent no efficacious or definite agreement is
possible. But time presses, and we must pass over
details. We prefer to leave them to your good faith,

and to the lively sympathy which you have often
expressed for our cause and its destinies. It is not,
we can assure you, by diplomatic means that we can
come to a mutual agreement, but by an appeal from
people to people, frankly and cordially expressed, without mistrust as without afterthought. France, more
than any other nation, is capable of hearing and
understanding this appeal. This appeal for the cessation of a normal state of things, and one which, between the French Republic and us, especially after
the declaration of your Assembly and the newly expressed sympathies of the French people for us, would,
if continued, become absolutely incomprehensible, we
now address to you for the last time, with all the
force of conviction and desire which is in us. May
you hold it sacred, sir, for it sums up the immovable
convictions and ardent desires of a small but brave
and honourable people, which has not forgotten its
ancestors, which has not forgotten that they did something for the world, and which, fighting to-day for a
sacred cause, that of its independence and liberty, is
irrevocably resolved to follow in their footsteps. This
people, sir, has the right to be understood by France,
and to find in her a prop and not a hostile power.
It has the right to expect from France *fraternity* and
not *protection*, the demand for which would, at the
present hour, be interpreted in Europe as a cry of distress, lowering it in its own eyes, and rendering it
unworthy of the protection of France, upon which it

has always relied. Such a cry of distress ill becomes us. There is no such a thing as impotence for a people which knows how to die, and it would be ungenerous on the part of a great and proud nation to misinterpret the sentiment which inspires us.

" It is time, sir, that this state of things ended; it is time that fraternity ceased to be an idle word, with no practical results; it is time that our messengers, our troops, and our arms should be able to circulate without let or hindrance throughout the length and breadth of our territory. It is time that the Romans should no longer have to regard with suspicion the men whom they have been accustomed to treat as friends. It is time that we should be free to defend ourselves, with all our resources, against the Austrians who are bombarding our city. It is time that there should be no mistake anywhere as to the good and loyal intentions of France. It is time that Europe should no longer be able to say that she deprives us of our means of defence, in order to force upon us by-and-bye a protection which would preserve for us our territorial integrity without preserving for us what is dearer to us by far, our honour and our liberty.

" Do that, sir, and many difficulties will be smoothed over, many sympathetic ties, now loosened, will be tightened once more, and France will have asserted her right of counsel amongst us far more effectual than by the apparent state of hostility which now subsists between us.

"The cantonments which, as well as we can judge, would be the most suitable, would be those upon the line which extends from Frascati to Villetri. The preamble of the declaration may very well be adopted as it stands.

(Signed)   "The Triumvirs,
            " ARMELLINI, J. MAZZINI, and SAFFI."

The General in command wrote to me declining to co-operate. He did not communicate his fresh instructions to me, this being evidently the result of the change which had occurred in the policy of our Government, but he wrote to me on the evening of May 31st :—

"Your convention is opposed to the instructions which I have received. I will not give it my assent, and I am obliged to declare this to the Roman authorities. When the Ministry, after having received the communications conveyed to it through M. de la Tour d'Auvergne, shall have made known its intentions, I shall conform scrupulously to them. In the meanwhile, I regret that I cannot possibly act in concert with you.

(Signed)   " OUDINOT."

My reply was as follows :—

"ROME, *June* 1, 1849.

" Monsieur le Général en Chef,—I have devotedly, and with much personal abnegation, followed the

directions which I received from the Government
of the Republic. Upon the day when, in presence of
many eye-witnesses, you created a scandalous scene
which nothing but my *sang-froid* and calm resolve
prevented from degenerating into personal violence,
upon the day when, ignoring me altogether, you gave
your commanders of corps orders to commence hostilities all of a sudden and under cover of the night, my
mind was made up beyond the possibility of change.
I left with you the day before yesterday, at 8 A.M.,
and at 3 P.M., and yesterday at 6 A.M., three notes,
copies of which I am also forwarding to the Minister
of Foreign Affairs. These documents will show that,
anticipating your projects, I had called upon you not
to put them into execution. You imagined that as I
had addressed an ultimatum to the Roman authorities,
the declaration I made to you that my mission was at
an end and that hostilities might recommence at the
expiry of the time named, was absolute and independent of any fresh incident which might arise.
But I stated to you in due time, and I here repeat,
that nine hours before the time named (twenty-four
hours) had expired the Roman authorities had replied
to our ultimatum ; that they had sent me a counter
project, which the plainest common sense, the elemen-
tary principles of diplomacy, and, above all, the
dictates of humanity, made it incumbent for us to take
into consideration.

"You could scarcely find time to cast your eyes

over this paper, or over the letters of the Municipality
of Rome, the President of the Constituent Assembly,
and of the Roman Executive Power. You sent me
back the correspondence by your chief aide-de-camp,
M. d'Espivent, who told me that you were too busy
to read it carefully ; you afterwards convoked Generals
Vaillant, Regnault St. Jean d'Angely, Mollière, the
chief commissioner of the army, and the chief of your
staff, Colonel de Tirion. In their presence, and notwithstanding your loud talk and threatening gestures,
I quietly read all my documents as well as my notes
of the day which I had addressed to you. As my
representations were useless, and as I formally refused
to associate myself with your project for a night
attack without previously warning the Roman authorities—an act which would have perhaps led to the
massacre of the French colony in Rome—I withdrew.
I am desirous of placing on record here the fact that
all the persons present at the meeting treated the
official representative of the Republic with the utmost
courtesy.

"Upon reflection, and in compliance with the
urgent and enlightened advice tendered you, you
decided at the eleventh hour to recall your order for
the resumption of hostilities. But these orders did
not arrive in time to prevent the occupation of Monte
Mario, where you met with no resistance, because I
had had time to let the authorities in Rome know
through my secretary, M. Leduc, that there was no

cause for alarm, your movements being merely made
to anticipate the foreign armies which were marching
on Rome from seizing these positions. But for this and
for my return to Rome the tocsin would have been
sounded, the garrison and the population of the city,
down even to the women of the Trastevere armed
with their knives, would have mounted to the assault
of Monte Mario. I feel sure that our brave soldiers
would have held their own, but the consequences of
an attack and of a desperate assault would have gone
straight to the heart of our country.

" Having left head-quarters after I had handed you
my last note, and having my eyes thoroughly open to
the objections to the immediate entrance of the French
army into Rome, where you would probably have compromised the interests which it is my duty to forward,
I drew up of myself a fresh draft of agreement, entirely
in conformity with the directions which I had received
from the French Government. This project, adopted
after some discussion by the executive power, has been
approved, with only three dissenting votes, by the
Constituent Assembly.

" I handed to you, before signing it, a copy accompanied by a declaration, first communicating to you
my instructions of May 8th.

" With regard to your declaration that you will
consider as null and of no effect the arrangement duly
signed yesterday by the executive power and myself,
it will be for our Government to decide, and, in accord-

ing with usage, you will not be at liberty to infringe it
in any way pending its ratification or non-ratification.

(Signed)   " F. DE LESSEPS.

" P.S.—The Triumvirate has forwarded to me a
copy of the letter which you addressed them this
morning with their reply. The step which you have
taken is a deplorable one, because it makes public a
difference of opinion as to which our Government was
the sole judge, and which for the present should have
remained a private matter between ourselves."

General Oudinot to the Triumvirs.

" *May* 31, 1849.

" Gentlemen,—I had the honour of informing you
this morning that I accepted, so far as regarded myself,
the ultimatum which was transmitted to you on the
29th by M. de Lesseps.

" To my great surprise, M. de Lesseps, upon his
return from Rome, brought me a species of convention
completely opposed to the basis and spirit of the ultimatum. I am convinced that in signing it he has
exceeded his powers.

" The instructions which I have received from my
Government formally prevent me from associating
myself with this last proceeding, which I regard of no
effect, and it is my duty to inform you of this without
delay.

(Signed)   " OUDINOT DE REGGIO."

The Triumvirs to General Oudinot.

"ROME, *June* 1.

"Monsieur le Général,—We have this moment received, with surprise and regret, your despatch of May 31st.

"The difference of opinion between the General in command and the Minister Plenipotentiary of France was not an event for which we could be prepared; and as this difference of opinion arises with regard to a convention the spirit of which is in entire harmony with the explicit aspirations which recently emanated from the French Assembly and with the well-grounded sympathies of your nation, it is a very deplorable occurrence, and one which may result in the gravest consequences, the responsibility for which does not rest with us.

"We hope that in the material interests of Rome, as in the moral interests of France, this difference of opinion will speedily disappear.

"For the Triumvirate,
            (Signed)   "JOSEPH MAZZINI."

I replied as follows to the Triumvirate:—

"*June* 1, 10 A.M.

"Gentlemen,—In reply to yours of this morning, containing General Oudinot's letter and your reply, I have the honour to inform you that I adhere to the arrangement signed yesterday, and that I am starting

for Paris in order to get it ratified. This arrangement was concluded by virtue of the instructions
which charged me to devote myself "exclusively
to the negotiations and relations which it might be
desirable to establish with the Roman authorities and
people."

(Signed)   "F. DE LESSEPS."

I consequently informed the Minister of Foreign
affairs that, in consequence of what had occurred, my
position would be a false one at head-quarters, that
it would be equally so at Rome, and that I must
therefore consider my mission as ended, or at all
events as in abeyance.

I was already making my preparations for a start
when M. de Gerando, the Chancellor of the French
Embassy at Rome, handed me in an open envelope,
upon behalf of the chief of the staff of the French
army, a telegraphic despatch thus worded from the
Minister of Foreign Affairs : —

"PARIS, *May* 29, 1849, 4 P.M.

"The Government of the Republic has put a stop
to your mission. You will be good enough to start
upon your return to France as soon as you have
received this despatch."

The orders of the Government found me ready to

execute them without delay, and, leaving Rome at
3 P.M., June 1st, I reached Paris at five in the morning of the 5th.

M. de Tocqueville had succeeded M. Drouyn de
Lhuys, and upon my calling to see him at his own
residence, he told me that he had not had time to read
my correspondence, and that he was not very well up
in the Roman question. I placed myself at his disposal for all the information which he might desire.
Neither M. Odilon Barrot, who before my departure
had said to me, at the time of the Assembly passing
the resolution which condemned the attack upon
Rome, " You will have to get us out of this dilemma,
we reckon on you," nor any member of the Government had expressed a desire to hear what I had to
say. Silence was necessary for them, in order to get
through the interval between the National and the
Legislative Assembly, in which the reactionary party
was to have a majority, and the Prince-President, who
was already preparing for his *coup d'etat*, received
me with his customary good nature, his sympathy for
Italy, and a show of indifference for our internal
situation. He none the less countersigned, in order
to maintain the silence which had been agreed upon
concerning Rome, the decree for the examination of
my conduct by the Council of State, by virtue of Art.
99 of the Constitution, which subjected to the jurisdiction of that body the high functionaries of the State
in certain predetermined cases, of which mine was the

only one. This article was not included in the future constitutions.

The Government never informed me of what they had to complain in my conduct; but M. Odilon Barrot, who made everything subservient to his keeping himself in power, completely eluded the question upon which he had to give a reply. For what had he to answer, if it was not to explain the sudden recall of the agent charged with negotiating with the Romans, and the order for our troops to enter Rome by force.

Instead of giving the hitherto concealed motives for these two decisions arrived at on May 29th, the President of the Council diverted the debate into another channel, and in order to escape from embarrassing questions he attacked my treaty of the 31st, which was not concluded until two days after the order for my recall. It must be remembered that the telegraphic despatch of May 29th, containing the order for my recall, reached head-quarters twelve hours after the signature of my convention, and that General Oudinot could not have availed himself of it, as falsely asserted by M. Odilon Barrot to justify his refusal to adhere to the convention. Nor must it be forgotten that the date of May 29th marked the change from the Constituent to the Legislative Assembly.

———

It appears necessary for me, in the interests of

historical truth, to sum up and refute the principal heads of complaint urged by M. Odilon Barrot.

1st. The provisional arrangement of May 31st compromised the honour of France and the dignity of our arms. M. Barrot blamed me for having placed France in a compromising position, and this in face of the statement of facts which I have just made—viz., that I was careful from first to last to reserve the full liberty of the Government, and that I scrupulously avoided pledging it to a share in the work which I believed to be in the interests of France both as regards the present and the future.

The treaty of May 29th only became binding if ratified by the Government. If I had been over-zealous it was easy to refuse ratification.

But in what respect was this treaty open to the severe criticism to which it was subjected? I need not make any secret now, as I had to do forty years ago, of the unvarying idea which served as a rule to all my actions, in the discharge of a mission which I had not solicited.

When, on leaving for Rome, I had put into my hands, as my main guiding point, the *Moniteur* of May 8th, which contained the resolution voted by the Assembly, I supposed—and I could not but suppose— that I was sent to carry out the wishes of the Assembly.

When the Minister of Foreign Affairs and the Prince President of the Republic gave me their final verbal instructions, there was nothing to show (nor if there had been should I have consented to do so) that I was not meant to take into serious account the resolutions of a sovereign authority, resolutions which I had heard discussed in advance, and the spirit of which was perfectly familiar to me. I may take this opportunity of adding that, from the time of my departure till my recall, I received no fresh instructions, nor a word in reply to the telegrams in which I asked for a simple *yes* or *no* as to the measures which I suggested—nothing, in short, which could modify the inspiration by which my course of conduct was guided.

Was I sent to Rome to insist upon the Romans opening the gates to our army, under pain, in the event of their refusal, of seeing their houses devastated and their fellow-citizens decimated by the sword ? It is evident that I was sent to come to an understanding with a population which we regarded as having lost its head, to bring together parties which, irritated by recent occurrences, could not act of themselves. There was no suggestion that I should facilitate a surprise or provoke a struggle, but what I had to do was to give the Romans a proof of our disinterestedness and friendship. I went to Rome to make the inhabitants feel that they would do well to place themselves under the protection of France and escape from all the consequences of reaction by accepting our support. The honour of our flag was

not dependent upon our occupying Rome at a given
date and at the hour we saw proper to fix; but what
we had to do was to be on the watch for, and if
necessary prevent, the entry of foreign troops, and be
ready to succour a friendly people should danger
threaten it. Can the honour of a nation like France
be banished because she treats considerately a city
which she wishes to take under her protection?

In the state in which I found Rome there were
two courses open, either to have recourse to force,
trampling my instructions under foot and being
untrue to the national will; or to do as I did, and,
by standing in the way of an imminent conflict, arrest
the unfortunate events which have since occurred
there.

2nd. The negotiations ought not to have been
resumed after the collective declaration of May 30th.
I think that I have explained clearly enough in the
above narrative of facts how I was led to resume
negotiations after the reception of another project
presented within the dates fixed by the ultimatum,
and how the drafting of my agreement, which was the
logical outcome of my instructions and of the circumstances in which I was placed, seemed to me to meet
the difficulties which met me on every side.

It may not be out of place to mention here that
the telegraphic order for my recall of May 29th had
not reached General Oudinot on the 30th, and that
the cause of it could not have been this treaty, which

the Ministry did not know of until the 6th of June.

3rd. My negotiations had facilitated the re-victualling of the city. From the beginning of the armistice until the 1st of June, General Oudinot did not allow the city to be revictualled. Communication was free at the points occupied by the French posts only for unarmed persons provided with proper safe-conducts and for small quantities of provisions. Several French merchants sent me a petition asking me to authorise their fetching from Civita Vecchia season goods over which they would incur serious loss if their sale was delayed. I communicated their petition to the General, who refused to do as they asked.

4th. I am accused of having unnecessarily kept the troops inactive. It is quite clear that my mission as a negotiator did not admit of my engaging in hostilities, and that if the troops had been ready to attack they would have had to wait. But I maintain that they were not in a position to undertake the siege when I reached head-quarters on the 15th of May, which reduces the period of my negotiations to a fortnight and not to a month as has been stated. During this time the French troops did not remain inactive, and the preparations for a siege were not suspended for an hour. A large number of fascines were made every day, and movements which excited the apprehensions of the Roman populations and often hampered me in my intercourse with the authorities were constantly being carried out. The bridge of boats on the Lower Tiber was being got ready, and it was thrown across the river previous to the rupture of negotiations, though it interrupted all communication between Rome and the sea by water, and though, by closing the passage to fishermen's boats, it deprived a portion of the population of their means of livelihood.

A *tête-de-pont* was being constructed on the left bank of the Tiber, and troops were sent there, although this point had not been occupied at the commencement of the armistice; the large church and convent of St. Paul, which are still nearer to the city, were also seized. Thus the time which I spent in negotiating was not lost to the army, and, what is still more to the point, the reinforcements which the Government had on the 10th of May telegraphed to General Oudinot to await had not yet arrived when I reached Civita Vecchia. A letter of June 18th gives this explanation of the delay in the siege which assuredly could not be attributed to me after my departure.

5th. It is said that the armistice gave time to all the men who disturb Italy to assemble in Rome and form an army which now confronts us.

" The forces which were defending Rome did not increase during my stay there as alleged by the President of the Council. At the end of my negotiations they were the same as they had been at the beginning, and if the delay in attacking has been of advantage to either side, it has notoriously been so to the French forces.

Not a single foreigner joined the Roman army during my stay, and those who had been there before I came consisted simply of some twenty Frenchmen, a few Germans, and from 150 to 200 Poles, who expressed to me *in writing* their desire to quit Rome rather than join in hostilities against France, and to go wherever we might see fit to send them.

With regard to the Italians of states other than the Romagna, are they to be considered as strangers to the cause for which Rome is struggling? In any event, it would be absurd to attach much importance to them in a city which contains 30,000 regular soldiers, and a whole population in arms ready to offer the most desperate resistance. I had already informed the Government of these facts by a despatch dated May 16th, and I had specially instructed MM. de Forbin-Janson and de la Tour d'Auvergne to confirm their tenour.

6th. I should have stipulated for the occupation of Rome, as the only means of holding a high tone with the foreign armies which were advancing.

Is it fair, is it reasonable to reproach me with not having insisted, as the *sine quâ non* of any agreement, upon a clause for the occupation of Rome, when M. Drouyn de Lhuys declared in the sitting of

THE MISSION TO ROME. 97

May 7th, not only that he had not given orders for the attack upon Rome, but that he "had only authorised the march upon Rome on the condition of no serious resistance being offered, or of our being appealed to by the population at large"?

Can there be any more flagrant contradiction than that which is involved in this utterance, and the order given to attack and seize Rome before the result of our negotiations could be known?

Lastly, was it possible for us to have mingled without restriction with the Roman population and garrison, while preserving a mixed and expectant attitude, in conformity with the object of the expedition and that of my mission?

The permanent occupation of Rome by our troops was not indicated either directly or indirectly by my instructions as an indispensable element of the conciliation which I was instructed to bring about. It exposed us to countless difficulties. The Roman authorities have incessantly declared in their notes, as I have pointed out, that they could not agree to it so long as we had not recognised their Republic and the powers by which it was governed.

As to the "firm and resolute" language which our army was to have employed, when it had once taken possession of Rome, I do not know upon what such an idea is founded. If we had entered Rome after having destroyed the Republic, we should have had no need to have employed that tone to any one, for we should have begun by doing what the Austrians, Spaniards, and Neapolitans would have liked to do. If, upon the other hand, we had made our entry under the cover of treaties, and in promising to maintain a National Government of some kind (either the old or the new), to uphold the laws of the country, and respect the free will of the inhabitants, I would ask whether war with Austria might not have resulted from a situation of this kind, in the event of the Imperial troops advancing beneath the walls of Rome occupied by our army and manifesting their intention of restoring the temporal power of the Pope upon the lines indicated by the Court of Gaëta.

The outer cantonments of the city occupying healthy and strong positions, the possibility of the General in command residing upon the French properties at Monte Pincio, and bringing into the city, one after the other, the whole of his army corps, which indeed the population would have clamoured for the day my convention was signed ; these, surely, were conditions which satisfied the aim of the expedition, while satisfying the military honour of the army and the *amour propre* of the General. What has an opposite policy done for us, and what embarrassments has it not in store for us ?

7th. The last project agreed to by General Oudinot would have been met with the jeers and murmurings of the Roman Assembly, so that I ought not to have pursued the negotiations further.

It is not accurate to say that General Oudinot's project would have been so received, and if it had been I should have broken off the negotiations instanter.

I have more than once shown, in the course of twenty-three years' service abroad, that I am neither patient nor slow of speech when the honour and dignity of my country are concerned. .

*   *   *   *   *

I have mentioned above, in reply to complaint No. 5, that I had specially charged MM. de Forbin-Janson and de la Tour d'Auvergne to communicate to the Government the exact position of the Roman army. With regard to the latter, he was at this time a débutant in diplomacy. I had taken a great liking for him on account of his distinction and ability, and I foretold that his career would be a brilliant one. I did not see him again for twenty years, when he was ambassador in London during the period of my struggle with Lord Palmerston on the Suez Canal question.* One evening after dinner he admitted to me that on his return to Paris from Rome M. Drouyn de Lhuys questioned him a good deal about me, and asked him if he had not noticed that I was rather flighty. He confessed, with some confusion, that he had answered in the affirmative, little suspecting to what use the

* Note of the Translator.—Twenty years after 1849 would bring us to 1869, the year in which the Suez Canal was opened, while Lord Palmerston died in 1865. But M. de Lesseps does not intend to be taken too literally.

Minister would put his answer, which was a perfectly innocent one, and related to a conversation which we had during one of our visits to the monuments of Rome. "You see," I said to him, "all these fine palaces which are now unoccupied: in twenty years' time they will be the refuge places of all the petty sovereigns of Italy." Brought up by his brother, the cardinal, in the ideas of the previous century, the young man had taken my remark as the sign of a diseased imagination. It was on this account, no doubt, that M. Drouyn de Lhuys, after his betrayal of me, endeavoured to make people believe that I was off my head. The story got about among the clerks of the Foreign Office, and upon my return to France it was mentioned in print by the *Siècle*. I at once wrote to the manager of that paper, M. Chambolle, to ask him what he meant, and he at once contradicted the report in very proper terms. I went at once to the Ministry of Foreign Affairs, and met M. Drouyn de Lhuys as he was leaving his room after handing over the conduct of affairs to M. de Tocqueville. I met him on the stairs, and as I looked him straight in the face he hastily ascended to the next story and entered the bureau occupied by my brother. After seeing M. de Tocqueville, who informed him that he had not yet made himself familiar with the Roman question, I proceeded to the Elysée, where the Prince President greeted me very kindly; but while on the day of my departure he spoke his mind freely as to the revolutionists who had embraced the same cause which he had served in his youth—viz. in 1831, when his elder brother died of fever under the walls of Rome —he this time spoke of the obstinacy of the Roman Court, to which he attributed the impossibility of anything like conciliation. He appeared to me very uneasy as to the position of his Government, placed between the cross fire of two irreconcilable parties—the Reactionists and the party of the Mountain under Ledru-Rollin. I felt convinced from that hour that the President was preparing to combat them both.

I was not, therefore, surprised when I saw soon after the walls of Paris covered with bills announcing the invasion of the Chamber by the troops, the flight of Ledru-Rollin after the affray at the Arts-et-Métiers, and the arrest of several deputies.

Holding aloof, as I have always done, from the internal dissensions of my country, I awaited quietly at home the decisions which might be come to with regard to myself, and the *Moniteur* shortly afterwards published a decree which, by virtue of Art. 99 of the Constitution, referred to the Council of the State the examination of my mission to Rome.

The Ministry which had entrusted me, in a very critical phase of its own existence, with a mission bristling with difficulties, and which had so readily abandoned me without deigning even to examine my action, had not only countenanced attacks upon me, but had itself attacked me from the national tribune while I maintained a complete silence, and before the Council of the State had had time to commence its work. I had only at the last moment availed myself of my right of defence, and I had done so in a memorandum addressed to the Council of State. I had spoken with the reserve, moderation, and sincerity which becomes a man who, out of respect for himself and for public opinion, does not choose to imitate his enemies. But it was thought that even thus I was taking too much upon myself, and my independence was denounced as an infringement of discipline, while as the plain statement of facts stripped bare the policy by which I had had the bad grace not to allow myself to be crushed, fresh blows were aimed at me during the sittings of the Legislative Assembly on the 6th and 7th of August.

At the sitting of the 7th, M. de Falloux endeavoured to cast doubts upon the value to be attached to my statement as to the nature of the resistance which Rome could offer, and as to the presence there of more than 25,000 regular soldiers. These figures, which I had given as far back as the 15th of May, the day of my arrival, had been communicated to me by General Oudinot after his entering the city.

I learnt that it was proposed, at whose instigation I knew not, to retard as much as possible, or perhaps adjourn *sine die*, my appearance before the Council. But on the 9th of July I wrote to M. Vivien, the President, demanding the execution of the decree, and he replied to me, but not till the 20th, that the legislative section was ready to hear the case, and he requested me to present my defence.

M. Boulay de la Meurthe, Vice-President of the Republic, after having made a semi-official effort to induce me to abandon the cause and failed, announced to me on the 28th that the legislative section would assemble on the 30th to hear my verbal explanations.

The sitting opened at midday. When President Vivien had explained the object of it, I asked him before entering upon the case to inform me if the Ministry had explained the motives which had induced it to submit the examination of my mission to the Council of State, and if the latter had set forth any special accusation of which I could be asked to clear myself. I said that I had thought that Art. 99 of the Constitution, adopted by virtue of the principle of the responsibility of Government servants, could only apply to my case if there existed some definite fact which involved my responsibility, and which was outside the competency of my natural judge, the Minister of Foreign Affairs. In that case, I could admit the competency of the Council of State, and though its forms of jurisdiction were not well defined, I should submit myself to its judgment in all confidence, in order to publicly crush all the calumnies spread about to my disadvantage, and to establish that the motives of my conduct were above all reproach. The answer made me was that the Council of State had not been called upon to give its opinion as to any allegation of wrong done; that it was merely instructed to examine my conduct, that it was not constituted into a tribunal, and that, so far as it was concerned, there was neither accuser nor accused. I did not insist any further, and merely observed that as the Minister of Foreign Affairs had not made any definite allegation of a kind to involve the responsibility of his agent to a tribunal, it seemed to me strange that he should have had recourse to the Council of State to decide whether I had discharged my duties well or badly. I added that, without disrespect to the councillors of the State, it might be held that they were not in a position to judge of a diplomatic negotiation, especially as they did not summon witnesses or hear counsel, and as they take no account of what they have styled "outward circumstances and external commentaries."

This matter having been discussed, the President read out the instructions which had been handed to me on the 8th of May by M. Drouyn de Lhuys. M. Vivien and myself both of us pointed out in the same breath, that in the copy communicated by the Ministry there was a phrase which was not to be found in the instructions which I have given at page 14 of this chapter. The phrase inserted was as follows: "Everything which will hasten the end of a *régime* destined by the force of events to perish."

I at once submitted to the Council the *original* of my instructions, signed by M. Drouyn de Lhuys,

and which I was following as the document communicated by the Minister was being read. The
members of the Council, to all of whom it was passed,
were able to see for themselves that this phrase was
not embodied in them. Thereupon I myself caused
the sitting to be adjourned, and wrote the next day
the following letter to the President of the Council
of State, M. Boulay de la Meurthe :—

" Paris, *July* 31, 1849.

" Monsieur le President,—On leaving the Council
of State yesterday, I went to the Ministry of Foreign
Affairs and ascertained from M. Viel-Castel, the director of the political department, that the minute of
the ministerial despatch of the 8th of May, containing
the instructions relative to my mission to Rome,
agreed in every detail with the copy which was
handed to me before my departure from Paris and
which I showed to the members of the legislative
section. I shall be obliged if you will communicate
this fact to the Councillors, who will form their
opinion as to the circumstances under which an important phrase was inserted in the copy certified and
communicated by the Ministry. This phrase was, to
my mind, very conclusive, as it might of itself alone
have served as a base to the system which was designed to prove that my conduct was inconsistent
with my instructions.

(Signed)   " F. DE LESSEPS."

At the second sitting of the Council of State, replying to an examination which lasted four hours, I
pointed out how impartially I had judged the internal
situation of Rome, free as I was from all political
preoccupation or influences. For, in truth, happening to be in Paris a very few days after my return
from Madrid, and being about to accept the legation
at Berne, I should not have agreed to undertake the
temporary mission to Italy unless I had had a well-
defined object placed before me, and if I had had to
deal all of a sudden with questions which I had not
had time to prepare myself for. All that I had, as I
considered, to do was to prevent a renewal of hostilities between the French army and the Romans,
and to avoid the recurrence of a misunderstanding
similar to that of April 30, which had created so
painful a sensation in France. To bring about a suspension of hostile demonstrations upon either side, to
prevent, pending further orders, a bloody collision
which neither the Ministry nor the National Assembly
then desired, to ascertain what fresh events had taken
place since April 30, to see that I did not involve or
allow any one else to involve my Government definitely either in war or peace until it had had time to
be informed of how things stood and could decide
for itself, and not to recognise but not to destroy
by force of arms the Roman Republic, such were the
points to which I was told to direct my attention
when I started from Paris.

As I stated to the Council, the Government was so far from intending to attack Rome with our troops, and was so anxious to follow a conciliatory course, that M. Drouyn de Lhuys himself introduced me in his own drawing-room to the envoy of the Roman Republic, Signor Accursi, a member of the Constituent Assembly, who had recently been Minister of the Interior under the Triumvirate. He proposed that we should travel together to Toulon, a suggestion which I deemed it inadvisable to accept. It was then agreed that he should go to Toulon alone and embark upon the first vessel sailing for Civita Vecchia. Finally, during the last few days of my stay in Rome, I received a visit from an Italian who brought me a note in M. Drouyn de Lhuys's own handwriting, because, the note said, he was a friend of Mazzini and might help to effect a settlement. In order to let the Council of State know what impressions I had derived during the early part of my sojourn at Rome, and of the view, fully justified by events, which I took of the situation, I read to them extracts from my journal kept from day to day. One of these extracts, dated May 15—19, contained the following passage :—

"The city is in arms, barricades and defensive works are being erected in all directions. The resistance will be a very general one. The English Consul, who has resided in Rome for thirty years, has shown me his despatches to Lord Palmerston. He is of the same opinion as myself.

" The captain of an American man-of-war, who has visited all the defensive works, has told me that it would require twenty thousand men at the least to take Rome, and after a regular siege.

" Lord Napier and the captain of the *Bulldog*, an English war vessel, are of the same opinion."

*   *   *   *   *

I learnt that the day after I had appeared before the Council of State the Duc d'Harcourt, ambassador to the Pope, who had been called to give evidence, insisted that all diplomatic action would be made impossible if the conduct adopted in regard to me was to prevail, and that he expressed himself in the most favourable terms towards me, asking how any one could have thought of blaming me for occurrences of which he had been an eye-witness, and of which he was better able than any one else to form an opinion, though he and I were not always agreed. The evidence of so competent a witness was of great weight, but there is no reference to it in the report of the Council of State,* to which I made the following reply: —

" The theory of the infallibility of the instructions given to an agent, inaugurated by the report of the Council of State, upsets all the ideas hitherto current in diplomacy, converts an agent intrusted with a mission into an automaton deprived of all initiative,

* Note of the Translator.—M. de Lesseps does not give the report of the Council of State, though he states (see page 118) that it was adverse to him.

and rivets him to a chain which would prevent him from executing any movement in all the circumstances which had not been foreseen or literally explained by his Government.

"In my own case I still maintain, despite the opinion of the Council, that I have not acted contrary to the letter or spirit of my instructions; but before attempting to prove it, by challenging the fundamental errors of the report, I must take up the defence of the true principles, by contrasting the views held by M. Martens with the doctrine propounded in the following paragraph:—

"'The instructions of the Government are in no case to be attenuated, extended, or modified by the aid of outward circumstances or external commentaries not forming part of them; all the rules of hierarchy and of responsibility would be set at naught if this principle was not strictly followed, and the Council of State would be wanting in its duty if it did not scrupulously adhere to the same.'

"Not only did M. Drouyn de Lhuys himself take a contrary view (see his instructions of May 8th, at page 14), but M. Martens, in his 'Manuel Diplomatique,' vol. i. p. 131, says, 'Even when the course which a diplomatic agent is to follow and his political actions are traced for him in his instructions, and his duty obliges him to conform to them, there are, however, cases in which the orders he has received are such that the execution of them would produce an effect

opposed to the views of his superior, and detrimental
to the interests of his country. In a case of this kind,
and assuming that the diplomatic agent, having the
aim of his mission clearly before him, should be
thoroughly convinced that in obeying the orders he
had received he would be running directly counter to
this aim, he might, and perhaps ought, to take it upon
himself to suspend the execution of them, losing no
time in informing his Minister of what he has
done, and in giving reasons to justify his conduct.

" 'Moreover, his responsibility is not determined by
the concessions which he may make, nor by the exi-
gencies which he may insist upon, and the extent of
which is laid down for him in his instructions; his
main duty consists in doing what is best to the utmost
of his ability.'

" The reporter of the Council of State declares that
my letter of instructions should have been my sole
guide, and, while blaming me for putting my own in-
terpretation upon it, its meaning is so far from being
clear to him, when it has not the light of external
commentaries to guide him, that he has to preface it
with a preamble of the political intentions which he
attributes to the Government, and that he has to com-
ment upon them, to make extracts from them, and
recast them, so to speak, in order to draw from them
any positive meaning to support his own views. And
I am to be blamed for having regarded as serious and
binding the formal engagements entered into by the

Government, with the majority of a sovereign Assembly! But if the Minister for Foreign Affairs had at the time an *arrière-pensée,* which I will not even now do him the injustice of supposing, was I the man to accept, in view of a wretched personal interest, a mission the object of which was to do the very opposite of what my country had the right to expect of me after the public statements made from the tribune?

" My instructions authorised me in so many words to be ' guided by circumstances,' so how can it be argued that they limit me within the hard and fast lines of the despatch of May 8th? Are the subsequent utterances of M. Drouyn de Lhuys, M. Odilon Barrot, and the President of the Republic to go for nothing? Is the speech of M. Barrot (President of the Council) on the 9th of May, announcing my departure and the object of my mission in conformity with the vote of the 7th, of no value in the eyes of the Council? Then in that case the speech of the Minister of Foreign Affairs, delivered on May 22nd, should be erased from the official record of the Assembly, when he said: 'As to the Roman expedition, it has been the subject of two debates. The second was a very recent one; the Government explained the object of the expedition; the Assembly expressed its views and made known its decrees, and an agent was at once sent to Rome and to head-quarters. *He took with him as his instructions the report of the debate in this Assembly, and he has been instructed to shape his course in conformity with it.'*

" Thus, even excluding from consideration the un-questioned principles which I have just referred to, it has been admitted by the official declaration of the Minister who signed the despatch of May 8th, containing my letter of instructions, that this letter was not to be my sole rule of conduct, that the aim of my mission was collaterally indicated by the *outside circumstances*, such as those which occurred in the course of my mission to Rome, or by the *external commentaries*, such as the votes of April 16th and May 7th, and the ministerial undertakings bearing upon them. The nature of these undertakings is very clearly indicated by the speeches of M. Odilon Barrot, the President of the Council, both at the˙ sittings of April 6th and May 7th, and again on June 9th, upon all three of which occasions he explicitly declares that the French expedition under General Oudinot was not sent with any intention of attacking the Roman Republic, of entering the city by force, or of restoring the rule of the Pope.

" The reporter of the Council admits that the discus-sions in the National Assembly do not in any way invalidate the character of my instructions, but he nevertheless asserts that all I had to concern myself with was what related to the entry of the (French) troops into Rome and with the special conventions calculated to secure it.

" The pretended necessity of having Rome occupied, despite the opposition of the Roman Assembly, the authorities, and the inhabitants, was not so much as referred to in my instructions, and it was in opposition with the statements of the Minister in the Chamber. Be this as it may, the reporter, taking as his starting point the principle of an entry, by force if necessary, into Rome, encompasses me within the circle which he has seen fit to trace, and beyond the limits of which I was not, according to him, to have stepped.

" The report charges that my first proposals did not produce any immediate effect, and that they underwent modifications. Inasmuch as I was instructed to treat, and as in all negotiations there are several contracting parties whose interests are different from one another, I could not, from the very outset, force my own views upon those with whom I was treating. I could not but admit the presentation of counter projects, discuss them, and be led perhaps, either by force of conviction or by the urgency of circumstances, to make concessions.

" The reporter blames me for not having shown sufficient consideration for the susceptibilities of the Papal Court at Gaëta. This is a question which the Council of State could not possibly have sufficient data for discussing; and the Minister of Foreign Affairs, in laying it down in my instructions, had certainly no idea of fixing any particular limit. The

susceptibilities in question had been excited in regard
to us long before my mission by the very principle of
our expedition, which had been undertaken without
the Holy Father having first had it notified to him;
by the maintenance of the Italian tricolour, which we
had allowed to float side by side with ours at Civita
Vecchia until after the capture of Rome ; by the first
proclamations of General Oudinot; by the expulsion
from Civita Vecchia of the three commissioners appointed to represent there the interests of the Holy
See ; and by the telegraphic despatch addressed to
General Oudinot as well as to myself, and beginning,
' Inform the Romans that we do not intend to act with
the Neapolitans *against them.*' This despatch evoked
accusations of treachery against us from Gaëta as well
as from the staff of the King of Naples, who had
already arrived almost within sight of Rome, and who
lost no time in raising his camp and hurrying back to
the frontiers of his kingdom.   It will be seen that the
very principle of my mission, aggravated by circumstances for which personally I was in no degree
responsible, was a permanent cause of irritation at
Gaëta.   I could not take a step without incurring to
some extent the reproach which the report of the
Council seems to admit as more or less well-founded,
but there was no help for it.

"The report says that I expressly disobeyed my
instructions :—

" 1st. In lending myself to acts which gave the Roman authorities a moral sanction.

" 2nd. In placing myself at variance with MM d'Harcourt and de Rayneval.

" 3rd. In making arrangements which were not partial, inasmuch as I had only to concern myself with what related to the entry of the (French) troops into Rome, and the special conventions calculated to secure that end."

" To this I reply :—

" 1st. Was the Council of State in a measure to judge how far my action gave any moral force to the Roman authorities ? The Council could know no more as to this than as to the susceptibilities of Gaëta. So far as I am concerned, I am convinced that I have done nothing to deserve such an imputation, and the details given in my memorandum should have sufficed to show that this was so. But, after all, did not my instructions authorise me, in so many words, 'to devote myself exclusively to the negotiations and the relations which it might be desirable to establish with the Roman authorities and inhabitants, and to come to terms with the men at this moment (May 8th) invested with power in the Roman States. I have carefully avoided going beyond the line I have laid down for myself, and it is a well-ascertained fact I have not recognised the Republic, the name of which does not appear in any of my agreements. It is a principle of diplomacy that the relations between the various

Powers and the *de facto* authorities of a foreign country do not necessarily imply the recognition of that authority.

"2nd. My instructions directed me, *save in urgent circumstances*, to act in concert with MM. d'Harcourt and de Rayneval, but they did not make it compulsory upon me to be, at all times and upon all points, in agreement with them, nor to follow absolutely their advice if I deemed it opposed to the aim of my mission, which was different from theirs. MM. d'Harcourt and de Rayneval, whose competence is undeniable, had quite understood this themselves. They would only have had the right to be exacting in regard to me if they had obtained from the Holy See something in the shape of liberal declarations of policy, and if their efforts had not, as M. d'Harcourt so clearly foresaw that they would, been counteracted by the reactionary tendencies of the Papal Court. It was for each of us to give his opinion to the Government, with whom it remained to examine and settle the question in the last resort, and give its orders accordingly. Indeed, M. de Rayneval wrote to me on May 28th : 'You have appealed to the supreme judgment of the Government; it is only right to await its decision.' The Minister of Foreign Affairs took the same view, for when I asked him, on the eve of my departure from Paris, for explanations as to the passage in my instructions which bore upon my relations with MM. d'Harcourt and de Rayneval, he replied to

me, in the presence of M. de Viel-Castel, now my col-
league in the French Academy,* 'Send them duplicates
of your despatches.'

" 3rd. I have already said that there was not a
word in my instructions which bade me concern myself
with arranging 'special conventions calculated to
secure the entry (of the French troops) into Rome.'
Consequently, according to the system laid down by
M. Vivien, by which one should do nothing not
expressly laid down in the instructions, I ought not
to have proposed that our troops should enter Rome.
Nevertheless, I did so several times; and even in connection with the arrangement of May 31st, I have
shown in my memorandum how we might have
gained a very important position inside the city, and
how we should have been asked to come in a very
short time by the inhabitants themselves. Specially
authorised, quite independently of my instructions,
to conclude *partial arrangements* with the Roman
authorities, I abstained from touching upon the principal question, viz. that of the relations between the
Pope and the Romans. The Council of State regards
me as exclusively responsible for the first projects of
arrangement proposed in concert with General Oudinot,
inasmuch as in condemning these proposals it blames
me alone; but it is inconsistent to reproach me with
having signed the provisional agreement of May 31st,
in spite of the opposition of the General, whose

* M. de Viel-Castel died October 6th, 1887.

responsibility was not so deeply engaged as mine, and whose co-operation I was not bound to accept."

It will have been gathered by what I have said that the Council of State had not taken into consideration at all the circumstances which had led to my mission, or those in which I was placed during my mission; my correspondence with the Minister of Foreign Affairs; or the information which I supplied him with, and which gave him the opportunity of sketching for me the policy he deemed best; the absence of any reply, any order, instruction, or indication, from the time of my arrival in Italy till my departure from Civita Vecchia on June 1st; or of the change of policy which suddenly occurred in Paris on May 29th, when the Legislative succeeded the Constituent Assembly. By virtue of Article 99 of the Constitution of 1848, the Government remitted the examination of my conduct to the Council of State, and secured " for reasons of State " a vote of condemnation passed unanimously less the one independent voice of M. Pons, of the Hérault department.

The vote of blame was a fortunate one for me, as, returning to private life, I have since been absolved from it by my country, which has shown its confidence in me by generously placing at my disposal the means for carrying out two great undertakings conducive to its glory and to the progress of the whole human race.

PARIS, 1886.

CHAPTER II.

SELECTED at the outset of the Revolution by M. de Lamartine for the French Embassy in Spain, I was about to repair to Madrid, when I received an extract from a Spanish journal, in which it was said that the people of Paris, after having seized the Tuileries, had stolen the things left there by an Infanta of Spain. The Royal Family, on leaving the Tuileries, had left behind them all their most valuable effects, and among others the jewellery of the Spanish princess, who was the wife of the Duc de Montpensier. I accordingly asked M. de Lamartine to let me take possession of this jewellery. He told me that he had no power over the invaders of the Tuileries, who had erected barricades and would not allow any one to enter the palace, but he advised me to go and see M. Marrast, the Mayor of Paris. The latter, formerly a writer in the *National*, with whom I was well acquainted, said to me, "The fact is I do not in the least know the people who occupy the Tuileries, and I have no idea what their plans and intentions may

be. M. de Lamartine and myself are in a very tick-
lish position, which does not admit of our coming into
conflict with them ; but as you have made up your
mind to go there and parley with them, I will give
you a letter of introduction for their leader, if they
have one, in your quality of representative of the
Republic in Spain." He at once wrote a note, which
I regret not having kept, but which I can quote from
memory. It ran as follows : " M. de Lesseps is appointed Ambassador of the French Republic in Spain.
He would like to take with him the effects belonging
to the Spanish Infanta. As she is a foreigner, it would
be advisable to respect what property she left at the
Tuileries. I will be obliged, therefore, if you will
hand over to M. de Lesseps the articles which this
' young person' asks for." I went with this note to
the Echelle wicket gate, by way of the Rue du Louvre,
where I saw a number of men in their shirt sleeves,
very untidy, some of them wounded and wearing ban-
dages on their heads. They asked me what I wanted,
and I replied—

" I am the Ambassador of the French Republic in
Spain. There is a Spanish newspaper which says that
you have been robbing the Infanta of Spain."

They asked me if I believed them to be thieves,
and I begged them to take me to their leader, as I
had a letter for him from the Mayor of Paris. They
accompanied me to the part of the palace which is still
standing, and I was presented to M. St. Amand, a

captain in and wearing the uniform of the National Guard, who was in the grand saloon of the Duchess of Orleans. . . . I had with me the King's groom of the chambers, who had brought with him a list of all the articles which had been left behind by the Royal Family. M. St. Amand, assuming an air of great dignity and authority, observed, "This is a very long list," to which I at once replied, "But when it is a question of giving back what does not belong to one, there can be no question of much or little." Whereupon a common man joined in and said, "This gentleman is quite right."

The crowd closed in, and as I was about to be shown into the rooms where the effects claimed had been collected, a young man in a white blouse, with very delicate hands and features, pushed my elbow and said in a whisper, "Persevere in the same course; all these people are much better than they have credit for being." He was a M. de Montaut, a student of the Polytechnic School, who has since entered the corps of ponts et chaussées, and who was the first engineer attached to the Suez Canal, where I entrusted him with one of the divisions of the works. The people thereupon, without further reference to the captain of the National Guard, conducted me into one of the rooms on the ground floor, facing the Rue de Rivoli, where all the effects belonging to the Royal Family had been laid out on tables and ticketed, with as much order as in a curiosity shop. On looking over them, list in hand, I could see nothing of the jewellery, plate, or above all of a splendid album, the cover of which was enriched with precious stones, and which contained drawings by the leading French artists. It was a family present given to the Infanta upon her marriage. I was told that, " ragged as you see us, we stored all the most valuable articles into carts and slept upon them, taking the jewellery and plate the next morning to the Ministry of Finance, and the album to the National Library." I arranged with the young Polytechnician to have the whole taken to the Spanish Embassy, and gave a receipt for what was deposited in the Treasury and the Library, the transfer taking place without any difficulty.

After taking leave of M. de Lamartine and arming myself with letters of introduction for the authorities of the departments bordering on Spain, I went all along the frontier from Bayonne to Perpignan, in order to make sure that no revolutionary propaganda was being prepared for Spain, in accordance with the conditions which had been frankly accepted by M. de Lamartine and his secretary, M. Bastide.

We did not know what view other States would take of the revolution which had just occurred, and we had every interest to keep up friendly relations with Spain, as in the event of any difficulty with other Powers, this would dispense us from the necessity of keeping an army upon the frontier.

Having spent eight years in Andalusia and Catalonia during a period of disturbance and revolution, I was on excellent terms with the Royal Family, the Government, and the generals of different parties. After having satisfied myself that there was nothing to fear on the French side of the frontier, I went to fetch my family at Barcelona, and we started for Madrid by way of the Catalonian mountains, where civil war had prevailed for the last twenty years, and which were still occupied by bands of insurgents, Carlists, smugglers, and brigands. I was asked if I should like to have an escort, but it would have done more harm than good. So I started with my wife and children. Every now and then a group of horsemen would pull up at a certain distance from us, and one of them would come forward to the carriage and ask who I was. Upon my telling him he rode back to his companions, and they at once made off.

In this way I reached Saragossa, where M. Charles Valois, the first secretary of the Embassy, had preceded me, and thence I reached Madrid upon the day following a mutiny of two regiments, which Marshal Narvaez, Duke of Valencia, had succeeded in quelling, after seeing General Fulgosio, Governor of Madrid, and several officers of his staff killed at the beginning of the revolt. I had scarcely taken up my residence at the French Embassy when I was called up in the middle of the night and told that a lady, thickly veiled, was waiting in my study to see me. I came down at once and recognised her as soon as she had lowered her veil. She was the wife of General Moreno de las Peñas, and she told me that her husband had been denounced by the sergeants of the regiments which had mutinied as the leader of the movement. The court-martial had accordingly sentenced him to be shot within the twenty-four hours if he was captured. His wife came to implore me to assist him in escaping, as 1 had done once before at Barcelona, by embarking him on board a French man-of-war which conveyed him to France. I told her that the situation was a very different one here, in the centre of Spain, from what it was on the seaboard at Barcelona, but that I would see what could be done if she would come back later in the day. As soon as it was daylight I went to see my friend Narvaez, and was much surprised at finding him come to open the door himself, with a very disturbed look upon his face. I explained in as few words as possible the object of my early visit, and he told me that he was afraid, when he heard the bell, that the police had come to inform him of the capture of Moreno, and that the latter had been his companion at the military school and in the great defensive struggle of 1808. He would, therefore, have been much pained if he had been compelled to have him shot. I shook him heartily by the hand, and it was accordingly arranged that I should avail myself of the departure of a French family for Bayonne that same day by the mail-coach to get the General away with them. Orders were given to the police to repair to some place away from the square of the post-office whence the mails started, and the General arrived in disguise, carrying a trunk. Narvaez has been given such a character for cruelty that this story may perhaps cause astonishment, for it has even been related of him that, in reply to the confessor who, upon his deathbed, asked him if he had forgiven his enemies, replied, " I have no need to do so, for I have had them all shot." This story is an outrageous calumny, for I have known few more generous and kindly men. Narvaez was always ready to sacrifice his own life, either to defend his country from the stranger or to maintain order at home.

A few days after this I was told that Mdlle. Eugénie de Montijo, accompanied by her governess, was waiting to see me in the drawing-room, in order to speak to me on a very pressing and important matter. It appeared that, on hearing of the revolt at Madrid, and without waiting to know the result, a regiment stationed at Valencia had mutinied, but as the revolt was unsuccessful the authorities assembled a court-martial, and thirteen officers belonging to the leading families at Court had been sentenced to death. The captain-general of the province had remitted the sentence to Madrid to be countersigned by the President before carrying it out, and the sister of one of the officers had come to implore of Mdlle. de Montijo, whose mother was grand mistress of the Court, to present her to Queen Isabella. She had taken her to the palace, and the young lady, after having implored the Queen's clemency, had fallen fainting at her feet. The Queen, deeply moved, had sent for the Prime Minister, who, however, was inflexible, declaring that he should be obliged, in the actual state of affairs, to resign if the sentence was not executed. But as the Queen had not yet affixed her signature to the death-warrant, she left the palace at Madrid and went to Aranjuez—a two hours journey—followed by all her Ministers.

It was at this juncture that my intervention was asked for. I could not well refuse it, but there seemed little if any hope of success. I sent for post horses, and during the journey went over everything which I could think of as likely to mollify the severe policy of Narvaez. I at last hit upon what seemed to be the best plan. On reaching the palace I waited in a gallery leading to the room where the Ministers had assembled, previous to submitting to the Queen the death-warrant for signature. I requested an usher to tell the Prime Minister that I wished to speak to him. He at once came out, and as we leant over the balustrade of the gallery I said quietly to him, "I have come to take leave of you, for you will see that, as the conditions of my mission to Spain were accepted by a sovereign Assembly because I might be able to exercise a salutary influence over your Government, if it is learnt that Mdlle de Montijo, belonging to one of the highest families in Spain, has unsuccessfully solicited my intervention to procure a pardon which, in my opinion, will strengthen rather than weaken you, there is nothing left for me but to retire and to take leave of you." Whereupon Narvaez, looking me straight in the face, and seeing how determined I was, shook me vigorously by the hand and said to me in Spanish, "You may be off, Ferdinand, with these men's heads in your pocket." I did not stop to hear more, and grasping Narvaez by the hand in turn, went back to Madrid, where I learnt that the Queen had, at the instance of Narvaez, signed the pardon of the condemned men.

A few days after this I received a message from the French Consul at Bilbao, informing me that a French merchant vessel, with forty-five political refugees on board, who had been implicated in some unsuccessful revolt, had left that port in the middle of the night, but had been obliged, owing to a violent tempest, to put back the next day. The authorities had laid an embargo upon the vessel, and had demanded that the refugees, who had embarked clandestinely, without passports, should be delivered up. The Consul had asked for a delay until he could communicate with me. I at once went to see Marshal Narvaez, and pointed out to him that we had no right to detain the Spanish refugees on board a merchant vessel, which did not enjoy the privilege of exterritoriality, which is exclusively reserved for men-of-war, and that the unhappy men were at his disposal. He did not hesitate a moment, and sent orders that the vessel should be allowed to leave with them for Bordeaux.

My personal position thus enabled me to render some service to my country during my mission at Madrid, and among other things I was able to conclude a postal convention, which had been under discussion for seventy years, and, while granting certain privileges to Spaniards, obtain the retrocession of the buildings of the church of St. Louis-des-Français, which had been under sequestration since the war of 1808.

Moreover, I had the satisfaction, when my mission was terminated by a change of residence which I had not solicited, of having the following words addressed to me by the Queen of Spain in public audience: "You carry away with you my esteem and that of all my subjects." If I repeat these words it is not out of personal vanity, but because they may be of some service to my country and to my children.

CHAPTER III.

A T the age of twenty I was sent upon a mission, in the year 1825, under the orders of my uncle, J. B. de Lesseps, the sole survivor of the Lapeyrouse expedition, who was then Chargé d'Affaires at Lisbon. Since then I have held different posts in the administration of Foreign Affairs at Tunis, in Algeria, in Egypt, in Holland, and in Spain. At the outbreak of the Revolution of 1848, M. de Lamartine summoned me from Barcelona to Paris, and sent me to Madrid as Minister Plenipotentiary. I had been eight years in Spain, during which time I had been upon terms of intimacy with the principal generals and public men, and though I had never mixed myself up in the political dissensions, I had established friendly relations with all the different party leaders. Lamartine said to me, " We are at the beginning of a revolution here ; we cannot tell if foreigners will be friendly to us. It is important for us that things should be quiet in Spain. You know the Court, the representatives of the different political parties, and the population at large ; and you have left a very good impression behind you. What I want you to do is to proceed to the Madrid Embassy, because, in the event of a foreign war, a good understanding with Spain is equivalent to 200,000 men on the Pyrenean frontier." I accordingly started for Madrid. Marshal Narvaez, who had no liking for the revolutionists, was in power, and I somewhat toned down his ardour and managed to save a certain number of persons who had compromised themselves. After a year's residence at Madrid, M. Drouyn de Lhuys saw fit to put Prince Napoleon in my place and to select me for the Legation at Berne. Upon the day of my arrival in Paris I went to the National Assembly and witnessed, from the diplomatists' gallery, a very stormy sitting. A telegram from Italy had just come in stating that General Oudinot, despite the declarations that had been publicly made, had attacked Italy, or rather the Roman Republic, and that the Government was gravely compromised. There was a talk even of sending the Prince President to Vincennes, of turning out the Ministry of course, and of giving strict injunctions for a complete change of policy. The irritation was very great in the Chamber, M. Ledru Rollin and the rest of the Extreme Left shaking their fists at the Ministry, and a free fight being imminent, when M. Senard, who was a man of considerable experience, calmed down his friends and got them to adjourn the sitting till the evening, in order to decide what should be done. During this interval the committees of the Chamber met, and M. Senard said to them, "The Government has acted very wrongly, but it has admitted the fact and has declared that it had given no orders, throwing all the responsibility upon the General. This being the case, if we despatch to Rome, without creating any crisis at home, a man upon whom we can rely, I feel convinced that the matter can be arranged." He then named me, and added, "I do not mean to say that he is a perfervid Republican, but he has always served his country well abroad without concerning himself with home politics, and if he accepts a mission he will carry it out faithfully."

I reached Rome, as explained in the first chapter, at a very critical moment. Garibaldi with his army was in the city, and knowing that the French troops would not interfere, he went in pursuit of the Neapolitan forces. When it was found that I was trying to effect an amicable arrangement the extreme party imagined that I was not acting loyally and determined to have their revenge upon me. I was informed of this project by a man of whom I shall have occasion to speak presently, one of those conspirators who are to be met with everywhere, who had been condemned to death in Spain but whose life I had saved. He was in turn to render me a like service, and in this wise. As soon as I reached Rome

I had summoned the Frenchmen in the city to meet
me, and informed them that I was about to commence
my negotiations, adding, " You will come and see
me again to-morrow. I will tell you what has been
done." They cheered me very heartily, and several
of them shook hands with me on the stairs. I had
with me General Vaillant, who was to have taken the
command in the event of an attack upon Rome, and to
have succeeded General Oudinot if we did not get on
together. The next day I was going, after a conference with the Triumvirs, to keep the appointment I
had made with the French residents, when a man
came rushing up to me with his hair flying in the
wind and exclaiming : " M. de Lesseps, I am in time,
as you have not started. Yesterday, when you came
down from the room where you had got the Frenchmen together, three men came close up to you. You
of course thought that they were your compatriots,
and one of them put out his hand. You took it, and
then turned round. Well, the man who shook hands
with you will do so again to-day, and then the one
beside him, who was watching your movements, will
cut your throat, as was done with Rossi." Rossi had
received a letter from a lady to whom the man who
warned me had written, informing her of the plot.
The letter was found in his pocket. There was no
absolute necessity for me to go to the Embassy, as I
could inform my compatriots by proxy of what had
been done, but I first made my informant swear that

any one whom I might send in my place would not
run any risk. He swore it upon the Gospel, but I
told him that as I knew he did not set any store
by that I must have a different sort of oath. He then
swore it upon the head of his sister, and I was satisfied. As M. de la Tour d'Auvergne, whom I had
sent in my stead, was some time coming back, I
began to get uneasy, when Prince Wolkonsky, the
Russian Chargé d'Affaires arrived, and said to me,
" When you assembled your compatriots yesterday (I
hope you will forgive me for what I am going to say,
but we are obliged to keep our Governments informed
of all that is important), I took advantage of my
familiarity with the Embassy during the time that
your predecessor, the Duc d'Harcourt, was there, to
make my way to a small staircase, the landing of
which is contiguous to the saloon in which you had
assembled your fellow-countrymen. I put my ear to
the partition and heard all that you said, and reported
it to my Government. I was about to do the same
thing to-day, when I heard three men speaking in
French. One of them said, 'Ah! the scoundrel has
not come to-day. If he had come, a few inches of
cold steel would have settled the job. Why did not
M. de Lesseps come?' "

One of my friends, Count Rampon, subsequently a
Vice-President of the Republican Senate in France,
and an old schoolfellow of mine, who was in the room
at the time, seized one of the men, and was going to

throw him into the street, but the Frenchmen surrounded him and pushed these three men into the
staircase where Prince Wolkonsky was standing. On
M. de la Tour d'Auvergne getting back, I asked him
what had occurred. He said, "Three men came up
to the carriage as I was starting and grumbled a little
because you had not come."

Colonel de Maubeuge had been sent to assist me in
my negotiations, so I despatched him to Mazzini to complain of this. I had the names of the three men, and
one was a Frenchman named Colin, who, like the two
others, has since died. I remember that on the previous night a dozen individuals had come and yelled
the " Ca ira " under the windows of the Embassy, so I
instructed Colonel de Maubeuge to inform Mazzini that
if the three men were not at once cast into prison I
should order General Oudinot to attack the city. He
replied that he had no power to do so. The man
who had saved my life was up to everything which
occurred, and it was arranged that he should generally
take his stand at the corner of a street facing the
hotel. I accordingly made a signal for him to come
and talk to me, and after informing him of Mazzini's
answer, asked what had best be done. He advised
my applying to Ciceronaccio, a popular leader of
great influence who had organised the revolution. So
I sent to tell Mazzini that if he could not calm the
population I must ask Ciceronaccio to do so, and the
effect was magical.

At nightfall I went out into the city to see what was being done, and found Garibaldi's army going off in pursuit of the Neapolitans. My anonymous adviser, who was at my side, urged me to see Mazzini that very night, and arranged to meet me in front of the Palace of the Consulta at one in the morning, at the foot of one of the great statues. I kept the appointment, and he then insisted that I should go up to the first floor of the palace, take off my boots, and steal past the soldiers on duty, who would probably be asleep, and find my way into the room at the further end of the palace, where I should find Mazzini fast asleep. This was rather a hardy and undiplomatic enterprise, but I undertook it, and reached the room where Mazzini was asleep. He had a very handsome face, I thought, as he lay there asleep; and though he had been exiled from so many States, he was then still a young man. I waited a little to see if he would awake, but as he did not I shouted his name. He jumped up in the bed, looked at me, and said, "Are you come to murder me?" I replied, "No, indeed; if one of us is to murder the other it will not be me. I have been told that you will not act openly with me. I have orders not to treat with you" (in consequence of the diplomatic difficulties which the fact of his being such a downright conspirator might have created for us with other States), "as it would not have done to let the world see that you held the thread of the negotiations I have come here to carry on. You have a Roman Assembly, composed of the great landowners in the country, who are devoted to your cause, and who are not mistrusted by Europe. It is with them that I must negotiate; but as you are the most important personage, I wish you to be kept informed of everything. You are aware that at the sitting which took place this evening before Garibaldi started it was decided that certain Roman statesmen, not, with one or two exceptions, residents in the city, were selected to negotiate with me, but you insisted upon being put in their place. Consequently, you have not kept your word or adhered to what was agreed upon between us." When a difficulty occurs a woman will burst into tears, but a man will throw himself into your arms.* Mazzini did this, and so we continued the negotiations. I found out afterwards that, urged on by his own party, he was somewhat opposed to the object of the negotiations, and that he kept up the agitation in Rome. I was at the head-quarters preparing for the negotiations, when Veyrassat, the man to whom I owed the previous information about Mazzini, came rushing into the camp, bathed in perspiration, and urged me not to go into Rome as I had arranged to do, in order to thank the Princess Beljioso and other ladies for the care they had taken of our wounded. But he urged me to alter my plans, as he said that the Piazza di Spagna and the Via Condotti were filled with people, and that the plot to assassinate me would

* Not in England, M. de Lesseps.   Note of the Translator.

indubitably be carried out. However, as all had been settled for my going, I asked at head-quarters for an officer to accompany me, and Commandant Espivent, who has recently been general in command at Marseilles, came with me in an open carriage. I proposed that we should each take two pistols, and that my servant, who rode in the rumble, should also carry two pistols to keep off those who might try to get up behind, as the men who use daggers are very afraid of firearms. I was therefore pretty sure that by acting boldly we should get through the crowd without any mishap. When we reached the Piazza di Spagna the horse of the gendarme who was riding in front of us stumbled, but fortunately did not fall. A petition had been presented to him calling upon the army to rise in rebellion, and he was imprudent enough to tear it up and throw it in the face of the public. The crowd then surged up to the carriage, but when they caught sight of my pistols they drew back. In this way our carriage reached the Hôtel d'Allemagne, and we all three walked backwards into the hotel so as to keep our faces to the crowd.

I resumed my negotiations with Mazzini, who seemed disposed to carry them out loyally; but I afterwards learnt that a Frenchman, who had accompanied Louis Napoleon in the Boulogne and Strasburg expeditions, was exciting him against the French army, and had advised him, in response to the present of an ambulance which we had made, to send our soldiers a number

of cigars, inside which were to be proclamations addressed to the French army.   I learnt the same day
from Veyrassat that Mazzini had on his table several
small sheets of very thin paper, upon which were
written the appeals to the French soldiers to mutiny.
He advised me to go to see Mazzini about two o'clock,
and as he had always many people to see him who
were generally on his right, I was to place myself on
his left, and I should then be able to lay hold of one
of these sheets, and prove to him that he had again
deceived me.

I did this, and was able to seize one of these pieces
of paper and put it in the crown of my hat.   I then
said to him, "Do you know what I am told. You were
twice led away by your friends, conspirators by habit,
and you have twice tried to deceive me.   This is the
third time.   I am informed that you have meditated
sending proclamations to the French troops. The French
soldier would burn down his mother's house if he received orders to do so.   Despite your experience, you
do not know the French soldier, and you have consequently made a great blunder. He denied the accusation.   Whereupon I said, taking the proclamation out
of my hat, "What do you mean by No? I have done
to-day a thing that I will never do again, and that is
to lay my hands upon this sheet of paper."   He then
again embraced me, and I followed up my negotiations,
which ended in a draft of agreement.

This over, I returned to Paris, and this is the origin

of the Suez Canal. The French Government had left
me to act without giving me a word of reply. I was
being deceived behind my back until the time came to
betray me openly. I sent my four secretaries of the
Embassy to Paris, but got no reply. Those in authority
were very well satisfied to compromise me, and it was
intended to react against the policy of the National
Assembly. Seeing this, I determined not to play a
double part. I was at last recalled, and when everything was ready the attack on Rome took place. I
accordingly returned to Paris. The Government tried
to make out that I was mad, and that has happened
once since. I could not stand this, and I resigned my
functions in the diplomatic service.

It was upon this that, having a very worthy
mother-in-law, who was as attached to me as I was
to her, a mother-in-law who had a large fortune while
I had none, I became her land agent. She owned
in the neighbourhood of Paris a property which was
of some value, but which involved a heavy expen-
diture; so I induced her to buy a large tract of
uncultivated land in the Berry district, and had it put
into cultivation. I built a model farm, which is still
in existence, and restored an ancient castle which had
belonged to Agnes Sorel.

While I was superintending all this, I learnt that
Abbas Pasha, the Viceroy of Egypt, was dead. He
was a very cruel and deceitful man, and had succeeded Mehemet Ali and Ibrahim Pasha in the

government of Egypt. His successor was the youngest son of Mehemet Ali, whom I had known well as a child, and taught to ride. He was enormously fat, and I made him take exercise, much to the delight of his father. This lad, who was very intelligent, was made to learn fourteen lessons a day. Mehemet Ali said to me one day, "As you are interested in my son, here are his notes." I told him that I did not wish to see them, as I could not read even then very well, and all I wanted to see was the last column showing his weight for the past and the present week. If there was an increase I should punish him, if there was a decrease I should reward him.

When I learnt my pupil's elevation to power, I wrote to congratulate him, and he replied, begging me to come and see him at once. As since my retirement I had studied in detail all the questions relating to the Suez Canal, I was perfectly familiar with the isthmus, and I was perfectly satisfied of the possibility of cutting the canal—an enterprise which had taken possession of my imagination after reading the memoirs of Lepère, the head engineer in the expedition of General Bonaparte.

I resumed my former investigations, being convinced that I should obtain the concession.

The Viceroy sent for me to come to Cairo, where he was about to assume possession of power, and he at once called together his generals to consult them on

the question.   As I rode out with them on horseback,
and as they were inclined to think more of a man who
could jump a fence than of a savant and a bookworm,
they were well disposed towards me ; and when the
Viceroy showed them the memorandum I had drawn
up, they were unanimous in my favour.   So I got my
concession, and this was the origin of the Suez Canal.

Once in possession of it, I said to the Viceroy, "I
am not a financier, or a man of business.   What do
you think I had best do?"   I had many colleagues
and friends who were rich.   I got a hundred of them
to join me, and proposed to found a company with
them.   We each of us put in a share of £200, and
this share is now worth over £40,000.   This sum
served for the preliminary investigations which I had
made by engineers whom I had brought from Europe
to examine the ground, which had never before been
done, as no one had ever dreamt that the canal could
be made except with the water of the Nile.   But I
had always been of opinion that as the two seas were
on the same level—stoutly as this was denied—the
work to be undertaken must be a purely maritime
one.   I stuck to my text in spite of all opposition, and
my obstinacy has had its reward.   I intend to act just
in the same way at Panama, though many engineers
would prefer, on account of the difference in level,
not of the seas, but of the tides, to construct a lock.
I would not have one at Suez, and I do not intend to
at Panama, as thus I effect a saving of more than a

million and a half. I accordingly appealed to my
friends, who each subscribed £200, and I went on
until all this was spent. I then said to the Viceroy,
" The question as to the possibility of making the
canal is settled. Would you like me to put myself in
the hands of financiers at Paris who will probably get
the best of me?" He replied that he had a good
reserve fund (Egyptian finance was not in the terrible
state that it is in now), and would bear all the cost.
And in forming my company I introduced a clause
according to which a certain percentage of the profits
was to go to the Egyptian Government. This being
settled, I set to work. We continued to make our
surveys for the canal, but the opposition of England
was at one time so pertinacious that the unhappy
Prince was at his wits' end. It was no use his
saying, " I have imprudently granted the concession
to a friend, a Frenchman; you must apply to him or
to his Government. I cannot withdraw it." The
English opposition did not disarm for that, and he
was positively wasting away, so I said to him one
day, "There is only one course left open. We will
continue our surveys for the canal, which is outside
Egypt, in the desert. But the fact will be known,
and you will be constantly pestered on the subject.
Let us take another line. There is a population in
the Soudan, which has been much oppressed by your
family. You have a brother who was massacred near
Khartoum." Mehemet Ali, I should explain, had

originally sent his brother-in-law into the Soudan with 100,000 men, and he had brought back with him the same number of slaves. He afterwards sent one of his sons to collect the tributes, or taxes, which his brother-in-law had levied upon the country. These taxes consisted of a thousand articles of each kind—viz., a thousand loads of straw, a thousand loads of wood, a thousand loads of corn, a thousand maidens, and a thousand male slaves ; all of this was brought in and placed in the camp. But the chiefs of the country formed a plot to destroy the camp, and at night, while the leaders of the force were celebrating their captures, they set fire to the wood and straw, so that not a man escaped alive. I accordingly advised the Viceroy to take only a few soldiers, besides myself, and to confer upon these populations just and beneficent laws.

We reached the frontiers of Egypt, near Korosko, and went on to Bou-Ahmed, on the confines of the Desert, having with us two caravans, which kept two days' march apart from each other, so as not to exhaust the water which was to be had on the road. At Bou-Ahmed I wished the Viceroy a " Happy new year," and in the evening rejoined him at Berber, which is close by. I found him in a terrible state of excitement, shedding copious tears, and when I asked him what was wrong, he told me that he was weeping over the misery which his family had wrought in the country. He said that since his arrival he had received petitions from every quarter, and had seen villages which had been burnt down and never rebuilt, adding that it was so sad a sight that he preferred to return at once to Egypt. I told him that he could not, enlightened ruler as he was, do this, and that it was his duty to give just laws to the inhabitants and introduce municipal institutions.

This cheered him a little, and we went on to Chendi, the very spot where his brother had been massacred. It is most remarkable how rapidly men can be got together in this country. You send out some messengers on dromedaries, and in the course of a few days you have assembled more than 100,000 men. I found a tent all ready for me at Chendi, and the Viceroy prepared me for a wonderful sight the next morning. And there, sure enough, were 100,000 men who had been collected in the space of three or four days, and whom he harangued as follows: "I have just learnt that the Turkish Sheik, who has governed this country for the last twenty years, has slaves confined at his residence, in disobedience of my orders. There is one slave chained up in his cellar. Go and release him." He then had the Sheik placed face to the ground before the assembled people, preparatory to flogging him and loading him with the chains which his slave had been wearing. This produced such an effect that the multitude shouted, "Allah! Allah! Long live the Khedive!" He then said to them, "You see those forts which my father had built forty years ago on the banks of the Nile to use against you. Go and take the cannon on their ramparts and throw them into the Nile." I whispered into his ear, " Perhaps, Monseigneur, you are going too far ; they may make use of them after you are gone." To which he rejoined, " They are quite worthless."

He was a shrewd politician. He went on to Khartoum, leaving the generals, ministers, and myself behind to register all the heads of families who had assembled. This took us about a day, our mode of procedure being to obtain the necessary information from the tribes who had representatives at Chendi. We placed so many numbers upon so many posts, and we each took down the names of the chiefs, of the wealthiest, the oldest, and the youngest. Having selected the leaders of the municipalities, we started on the following day to rejoin the Viceroy at Khartoum. Upon our arrival the Viceroy came out to meet me, and taking me by the hand said, " You must dine with me. You will hear such a band of music as never played before any sovereign—that of an ancient negro regiment which dates from my father's time. The regimental chemist has mended the wind instruments with soap plaster; and this was the band which welcomed me on my arrival."

Entering the dining-room, we had our dinner served on a small table placed beside the divan, and I noticed that towards the close of the meal the Viceroy's countenance began to cloud over. He had the habit when

he was put out to draw his red fez over his eyes, close
down upon his nose. He had the blood very much to
his head, and his neck and even his lips began to
swell, as if he was going to have an attack of apo-
plexy. What could be the matter ? All of a sudden
he got up, and unbelting his sword threw it to the
end of the long room, exclaiming, " Leave me alone !
Do not ask me what is the matter ! " We all left the
room, and he then sent for one of his confidants and
said, " Take M. de Lesseps to my room," which was a
magnificent one on the upper story ; and I could never
understand how at a place like Khartoum such splendid
furniture, tapestry, &c., could have been got together.

The Ministers were all in a great state of mind,
thinking that here, at five or six hundred leagues from
the capital, their Sovereign had suddenly gone mad.
We waited till two in the morning, but could get no
tidings except that his confidant told us then that he
had ordered a bath, no doubt to calm his nerves. I
mention this to show what Eastern princes of another
age were like. Absolute power has a tendency to drive
men mad. At three the following morning he sent for
me, and I found him quietly seated on a divan in a small
room smoking a pipe. He had calmed down, and he
said to me, " You have asked to take a turn upon the
White Nile and the Blue Nile. You have two boats and
my cook, so you can go on both these excursions." I
replied, " In other words, you send me about my
business. What was the matter with you last night,

and will you tell me ? " He had said to himself, " Here
is a man who leaves his family in Paris and comes all
the way to Khartoum to give me a piece of good
advice which had not occurred to me." This made
him so furious that he threw away his sword for fear
that he might forget himself and strike me with it.
He had known me since he was a child, and seeing
that his head was giving way he got rid of his weapon.
But he sent me away in order to be able to issue himself the grand decrees which have tranquillised the
country, and which restored to it a prosperity only
broken by the English expedition. When Gordon
was at Khartoum as governor the Viceroy informed
me that he had summoned him to Cairo to join the
Committee of Inquiry, of which I was president. I
said to him, "You are wrong. Gordon is a man of
great ability, very intelligent, very honest, and very
plucky, but he keeps all the Soudan accounts in his
pocket, written on small pieces of paper. All that he
pays out he puts in his right pocket, and all that he
receives in his left. He then makes up two bags and
sends them to Cairo, and money is sent back to him.
He is not the man to regulate the affairs of Egypt."
The Viceroy then telegraphed to him to remain where
he was, but he was so active that he came all the same,
as he was administering the country in a very able
manner, according to the traditions left by Mohammed
Saïd. I asked him to peruse the explanations of these
decrees, which he had not read, and which I had got

translated. He followed them up afterwards, and if there had been no English expedition the Soudan would not be, as it is now, a standing menace to Egypt. These are historical facts which I am stating, and which are not to be had elsewhere.

I thought that the opportunity was a favourable one for disclosing all that occurred in Egypt. Since then, being in London at the time of the English expedition, I learnt that Alexandria was about to be bombarded. No one else knew of it, so I came at once back to Paris and begged MM. de Freycinet and Ferry to come from the Elysée, where a Cabinet Council was being held. I said to them, " I warn you that Alexandria, which we have created, and which, thanks to the engineers, sailors, &c., whom we have sent out, has prospered, is about to be bombarded. Well, France must not be responsible for the carnage. When I knew it, it had a population of 45,000 ; now it has 200,000 inhabitants. It was created by France, and we cannot bombard it." A telegram was then sent to our fleet, ordering it to withdraw. I relate all these facts, as they are but little known. Our Government, which behaved very straightforwardly in the matter, quite understood the situation, and could have no part or lot in the destruction of the town.

The prosperity of Egypt dates from the expedition of Bonaparte and the arrival of the French in the country, and now it is on the high road to ruin. I do not scruple, when I am in England, to tell the English that they will never be able to do any good there. Since the beginning of history all the conquerors of Egypt have been obliged to abandon it : Persians, Assyrians, Greeks, and all. The reason is a very simple one. None of the Europeans or foreigners can reproduce their species there, and a country in which this is the case is one which cannot be permanently inhabited or governed. I hope the English will in the end see this. They have already lost a great many men, and they have had to abandon the Soudan. What is wanted is a better organisation. I regret, as I have more than once openly asserted, that the ex-Khedive was dethroned in favour of his son, who is a very worthy young man, but who has neither the power nor the authority of his predecessor, who had covered the country with telegraphs and railways, and who is, to my mind, the only ruler who can with advantage be sent back there. I do not go in for diplomacy, I give my opinions before all the world, and I declare that the only way to save Egypt is to restore, not the exclusive influence of France, but the influence which she has legitimately acquired by civilising the Egyptian people for the last fifty or even eighty years. France has no desire to lord it over other nations, but she desires to maintain the influence to which she is entitled.

It has been seen how we went to work at Suez, and it will be the same with Panama, and I hope with the same satisfactory results. We spent over twenty

millions on the Suez Canal, and we have given back
to France (as I showed in a memorandum handed by
me to the Government) fifty millions.

This is why I have so many backers among the
general public and common people. There is scarcely
a small tradesman or peasant who has not his share in
the Suez Canal Company. The other day I drove to
my office in a cab, and when I had given the driver
his thirty-five sous he took my hand and said, "M. de
Lesseps, I am one of your shareholders." These
are the men who made the Suez and will make the
Panama Canal, and Panama will be opened in 1889.
The example of Suez increases the number and confidence of the Panama shareholders. The Suez Canal
was opened sixteen years ago,* and we are still at
work upon it. At Panama we have means at our
disposal which we did not possess at Suez, and according to the estimate which has been made we have
57,000 horse-power, which at the rate of ten men
each represents an army of 570,000 men. Add to
them the 20,000 workmen on the ground, and it will
be seen that we shall be able, by 1889, to open a
passage sufficient for the purposes of navigation, while
after that we shall enlarge the canal, as we have done
for Suez, which yields such a magnificent return to
the shareholders, and upon which, nevertheless, we
are still at work. I have just returned from Panama,
where I went in the company of several distinguished

* This chapter was written in 1885.   Note of the Translator.

engineers and representatives of chambers of commerce, whose reports will be published and will tell their own tale. Frenchmen alone could do all this without the assistance of government or of capitalists; they are the most devoted and disinterested people in the world, and they made the Suez as they will make the Panama Canal. We have engineers of the greatest merit, men who are young; and I like young men, though I am myself an octogenarian. Old age foresees and youth acts. Well, we have in Panama five divisions of engineers, and every one is convinced that we cannot fail to attain the desired end. We have, from the Pacific to the Atlantic, a succession of workshops and sheds, with all the means which are now available for executing works of this kind. We saw whole mountains blown up with dynamite, blocks of stone measuring more than a hundred cubic yards sent up into the air like so many pebbles. We are delighted to be able to state, after the voyage which I have just made, that the canal will be open in 1889.

CHAPTER IV.

THE ORIGIN OF THE SUEZ CANAL.

To M. S. W. Ruyssenaers, Consul-General of Holland in Egypt.

"Paris, *July* 8, 1852.

"IT is now three years since I asked and obtained permission to be placed upon the retired list as Minister Plenipotentiary in consequence of what occurred in reference to my mission to Rome.

"Since that time I have been studying in all its different bearings a question which I had already been considering when we made acquaintance with each other in Egypt twenty years ago.

"I confess that my scheme is still in the clouds, and I do not conceal from myself that, as long as I am the only person who believes it to be possible, that is tantamount to saying it is impossible. What is wanting to make it acceptable to the public is a basis of some kind, and it is in order to obtain this basis that I seek your co-operation.

"I am referring to the piercing of the Isthmus of Suez, which has been talked of from the earliest historical times, and which, for that very reason, is regarded as impossible of execution. For we read, in fact, in the geographical dictionaries that the project would have been carried out long since if the obstacles to it had not been insurmountable.

" I send you a memorandum which embodies my ancient and more recent studies, and I have had it translated into Arabic by my friend Duchenoud, who is the best of the Government interpreters. This document is a very confidential one. You will form your own opinion as to whether the present Viceroy, Abbas Pasha, is the man to comprehend the benefit which this scheme would confer upon Egypt, and whether he would be disposed to aid in carrying it out."

To the same.

" PARIS, *November* 15, 1852.

" When you wrote me that there was no chance of getting Abbas Pasha to accept the idea of the piercing of the Isthmus of Suez I communicated my scheme to a financial friend, M. Benoit Fould, who was concerned in a scheme for founding a Crédit Mobilier at Constantinople. He was struck by the grandeur of the undertaking, and the advantages there would be in including, among the concessions to be applied for from Turkey, the privilege of executing the Suez Canal. The negotiator sent to Constantinople encountered difficulties which compelled him to abandon the project. One of the arguments used against him was the impossibility of taking the initiative of a work to be executed in Egypt, where the Viceroy alone had the right to decide what should be done.

"This being the case, I must shelve for a time my memorandum on the subject, and I am going to see about the construction of a model farm upon a property which my mother-in-law, Madame Delamalle, has recently purchased."

To the same.

"La Chénaie, *September* 15, 1854.

"I was busy with my masons and carpenters, who are building an additional story to the old manor-house of Agnes Sorel, when the postman appeared in the courtyard with the Paris letters. They were handed up to me by the workmen, and what was my surprise to learn of the death of Abbas Pasha, and the accession to power of our early friend, the intelligent and sympathetic Mohammed Said! I at once came down from the building, and lost not an hour in writing to the new Viceroy to congratulate him on his accession. I reminded him that the course of political events had left me idle, and that I should take advantage of my liberty to go and present him my homage, if he would let me know the time of his return from Constantinople, where he was to go for investiture.

"He replied to me at once, and fixed the beginning of November for me to meet him at Alexandria. I wish you to be one of the first to know that I shall be punctual in arrival. What a pleasure it will be to meet again upon the soil of Egypt, where we first came together ! Do not say a word about the piercing of the isthmus before I arrive."

To Madame Delamalle, Paris.

(Diary.)

"ALEXANDRIA, *November* 7, 1854.

" The Messageries steamer, the *Lycurgue,* landed me at eight this morning at Alexandria. My good friend Ruyssenaers and Hafouz Pasha, the Minister of Marine, came to meet me on behalf of the Viceroy, and I proceeded in a court carriage to one of his Highness's villas, about two and a half miles from Alexandria, on the Mahmoudie Canal.

" A whole battalion of servants was drawn up on the flight of stone steps, and they saluted me three times, putting out their right hands to the ground and then carrying them up to their foreheads. They were all Turks and Arabs, with the exception of a Greek valet and a Marseilles cook named Ferdinand.

" Here is a description of my house, which I remember having seen built many years ago by M. de Cerisey, the celebrated naval constructor, and the founder of the Alexandria arsenal, from which he launched in a very short time twelve vessels of the line and twelve frigates. M. de Cerisey contributed in no small measure, under Mehemet Ali, to the deliverance of Egypt from the many burdens put upon it. The principal pavilion is in the middle of a lovely garden, with two avenues of trees leading up to it, one from the plain of Alexandria, in the direction of the Gate of Rosetta, and the other from the Mahmoudie Canal. Up till the other day it was occupied by the princess who recently bore a son to Said Pasha, who bears the name of Toussoum. The reception-rooms and dining-room are on the ground floor, while on the first floor there is a very bright drawing-room, with four rich divans running round it, and with four large windows looking on to the two avenues. Leading out of it is the bedroom, with a very elaborate bed, the hangings of which are of handsome yellow silk embroidered with red flowers and gold fringe. Inside these there are double curtains of figured tulle. Communicating with the bedroom are two dressing-rooms, the first of which has rosewood and marble furniture, while in the second, which is equally elegant, the washing utensils are in silver, the soft towels being all embroidered in gold.

"I had scarcely completed the inspection of my apartments when several intimate friends of the Viceroy came to call upon me. I got them to tell about Said Pasha's habits since he came to the throne, what were his tastes and his tendencies, who were the people he had about him, who seemed for the time to be in favour or disfavour; in short, all that is desirable to know beforehand when you are the guest of a prince. They told me that since his return from Constantinople he had often alluded to my visit, and spoken to those about him of his former friendship for me. I was told that he had waited to take me with him to Cairo, a journey which he was intending to make by way of the desert, along the Libyan chain of mountains, with an army of ten thousand men. This journey will certainly be an interesting one, and will take ten or twelve days. The start is fixed for Sunday.

"I see that there is a fresh batch of servants just come in, viz. a *kaouadji* (chief coffee-maker), accompanied by several assistants and *chiboukchi bachi* (a chief of the pipes), escorted by four acolytes with their insignia, consisting of a dozen long pipes with large amber mouthpieces studded with diamonds. Nor is the mission of these personages a sinecure, for in the house of a great Turkish nobleman pipes and small cups of coffee (*findjanes*) are provided fresh for each caller.

"An officer of the Viceroy subsequently came to inform me that his Highness would receive me at twelve at the Gabbari Palace.

"I thought that from the very fact of my having known the Prince when he was in a very different position that it was all the more incumbent upon me to treat him with the respectful deference which is always so acceptable to the human heart. So I fastened on to my dress coat all my stars and orders. The Vice-

roy received me with great affection, speaking to me of
his early days, of how I had sometimes taken his part
when his father was very severe upon him, of the per-
secution to which he had been subjected during the
reign of Abbas Pasha, and, lastly, of his desire to do
what was right and make Egypt prosperous.   I congratulated him on his intentions, adding that if
Providence had intrusted the most absolute government on earth to a prince who had in his youth
received a most thorough education, and who later
in life had been severely tried by fortune, it was
for a great purpose, and that he would, I was convinced, justify his mission.

" We discussed the forthcoming military excursion
into the desert, and it was arranged that I should join
the party without having to make any preparations of
my own.

" When I returned to my pavilion at eleven in the
evening I found all my staff of servants drawn up in
the same order as before ; and the *chef* showed me a
very luxuriantly laid-out table, decorated with flowers,
and with several covers laid.   He said that orders had
been given for the table to be served in the same way
both morning and evening.   I told him that I should
only avail myself of this of a morning, and that I
intended to go to my bedroom.   Two footmen came
forward to help me mount the staircase, which was
brilliantly illuminated.   Just for once I allowed them
to do so, with all due gravity, as became the friend of

a sovereign, who ought to appear as if he was accustomed to receive similar homage."

" *November* 8, 1854.

" I get up at five. I open the two windows of my room, which are overhung by the branches of trees which I am not enough of a botanist to know by name. The air is perfumed with the flowers of these trees and of the jasmines which line the banks of the canal, beyond which, though the sun has not yet risen, is visible Lake Mareotis, its surface rippled by a light and pleasant breeze.

" I go then to pay an early visit to the Viceroy, who, as soon as he heard of my being there, came out of his apartments, and we recline on an easy divan placed in a gallery overlooking the garden. After we had enjoyed our pipes and coffee, the Viceroy takes me out on to the balcony of the gallery to show me one of his regiments of the guard, which is to escort him on his journey. We then go out into the garden to try some revolvers, which I have brought him from France.

" After our walk I tell him that I must leave him to go and receive at his house the persons whom I had invited in his name. He thanked me for doing the honours of *my* house so well.

" I pay a visit to my neighbour Halim Pasha, the Viceroy's brother. This young prince speaks French with ease and elegance. Fond of riding and shooting, he told me that he already found there was a double bond of

fraternity between us. He is to join us in the journey
through the desert, and will bring with him his
falcons and greyhounds. His arms and servants are
to be at my disposal."

" November 9, 1854.

" An early morning visit to the Viceroy, in his
father's palace of Raz-el-tyn, at the extremity of the
port. This is his divan for official receptions, and I
witnessed the first audience granted to the Consul-
General of Sardinia, who had to present his letters of
credit. After the audience we enter the private
apartments, where we have a long and very interesting conversation as to the best principles of government, but in the course of which not a word is said
about the Suez Canal, a subject which I do not intend
to broach till I am quite sure of my ground, and
until the question is so far ripe that the Prince may
adopt the idea as coming rather from himself than
from me.

" It is all the more necessary to act prudently as
Ruyssenaers remembers having heard him say, before
his accession to power, that his father, Mehemet Ali,
to whom the project of making the canal had been
suggested, had abandoned it upon account of the difficulties it might cause him with England ; and that if
he ever became Viceroy he should do as his father
had done.

" This precedent was not an encouraging one, but
I am convinced that I shall succeed."

"*November* 11, 1854.

"The Viceroy sends me a fine Arab horse, which he has had brought from Syria. I am informed that this morning there will be a review of troops in the plain between Alexandria and Lake Mareotis. I mount my charger and join the Viceroy, Soliman Pasha being in command of the troops. The soldiers go through their drill and practice firing, and as we are galloping along a diamond tassel drops from the Viceroy's cartridge-box, but he will not let us stop to pick it up."

"*November* 12, 1854.

"The Viceroy informs me that he is about to order his troops to commence the march to Cairo to-day, and he orders his aide-de-camp to take me early to-morrow to the first halt."

"*November* 13, 1854.

"I left the Viceroy's position at six this morning, riding the horse he made me a present of, followed by another led horse, two camels carrying my luggage, and accompanied by two cavasses mounted and two saïs on foot. We were to meet Zulfikar Pasha at the Gabarri Palace, and making the circuit of Lake Mareotis, regain the Viceroy's head-quarters. In order not to delay our march, our camels and luggage were placed in the charge of a cavass. After having left to our right the ancient baths of Cleopatra and the Arab's Tower we reached a well, around which the Viceroy had formed his encampment the night before. He had started at four in the morning to cross the lake at a point where it was almost dried up. As we followed the wheel-marks of his carriage, which had left deep ruts at the places where our horses' feet sank deep, we could see that the troops must have some difficulty in passing all the way. I talked to Zulfikar Pasha—whom I had known years before, when he was the companion of Mohammed Said in their boyhood—of my project, and understanding its importance for Egypt, he promised me that he would avail himself of his intimacy with the Viceroy to endeavour to pave the way for me, and make him favourable to the scheme.

" After crossing the lake we enter upon that part of the Libyan desert which, in ancient times, was an inhabited and civilised country, and which has, since the Arab conquest, been abandoned to a few Bedouin tribes. Every now and again one sees around the ancient wells some of those black tents, of camels' hair, which the Scriptures speak of, and which are still the same in Palestine, Syria, Arabia, and all the coast of Africa, from Egypt to Morocco.

" The sky clouds slightly over, and a slight breeze makes the atmosphere rather cooler than it was on the other side of the lake. I witness a regular desert scene : a dog is busily engaged in tearing to pieces a dead animal, and close beside him are solemnly stalking several birds of prey, which are awaiting their turn, and which do not stir at our approach.

" It is eleven o'clock, and Zulfikar and myself munch, as we ride along, the biscuits and sticks of chocolate which are very desirable substitutes for pistols in our saddle-bags.

" From an eminence we descry the Viceroy's camp. A Bedouin tells us that we shall reach it in half an hour ; but here, as everywhere else, the peasants have a way of misleading you as to distances, and I calculate that we have at least two hours before us. Not minding either the heat or the fatigue, I ride on without dismounting once. We reach the camp about half-past two, to find that the Viceroy is taking his rest, and that a tent has been prepared for Zulfikar and myself, next to his. Inside the tent I find an iron bedstead, with an excellent mattress, a counterpane of quilted silk, cocoa-nut matting, some folding chairs, and a mahogany table.

" The servants bring us pipes and coffee, followed by basins and ewers of silver, after which they sprinkle us with rose-water, by way of a preparation for our collation, which is brought to us on a salver placed upon a stool, around which we take our seats. I was about to use my fingers, like my companion, when a knife and fork were placed before me, they, like the spoons and plates, being of Sevres china. In conformity with the injunctions of the Prophet, there was no wine, but the iced water was excellent.

"The strains of a military band told us that the Viceroy was awake, and upon going out of our tent we met him coming from his. He called me in and explained to me how he had got his artillery across the lake, going from one battery to the other and urging on the men, for everyone had assured him that it was impossible for them to cross it. He was in the best of humours, and we spent a couple of hours discussing subjects, all of which interested much, and the main objective of which was that he desired to illustrate his accession to power by some great and useful enterprise. He listened to my remarks with much attention, and spoke without the slightest reserve, the time passing very quickly, and the ceremony of ablutions coming to warn us that it was the dinner hour.

" After dinner a courier arrived from Alexandria, with the despatches which had been sent by steamer from Constantinople, and he had them read over to him by Zulfikar, translating their contents to me as he read. These despatches were from his agent at Constantinople, and from Reschid Pasha, the Grand Vizier. Among them was one which he showed me, and which had been written by the Sultan's favourite in the harem, to thank him for a present of 150,000 piastres, which he had sent her. The letter also contained a message from the Sultan, complimenting him upon the appearance of the Egyptian troops which had recently been despatched to Turkey.

"The news from Sebastopol came up to the 2nd instant, and at that date the town had not been taken. The admirals had informed the generals that in another month's time it would be impossible to remain at sea, and this led to a decisive attack being made, which cost the allied armies from ten to fifteen thousand men.

"We are to remain here three days for two regiments of infantry, which are expected to-morrow, and for two cavalry regiments due the day after."

"*November* 14, 1854.

"I am up at five. The soldiers are just beginning to come out of their tents. The sky is cloudless, and the stars are still shining, while the moon lights up a vast plain which, despite its being so bare, is not devoid of charm. I go to wish the Viceroy good morning, and after having smoked a pipe and taken coffee together, we mount our horses and go to meet the two regiments. They soon come in sight, and they appear to be none the worse for their march, and in good condition. All that they had had since leaving Alexandria, early the previous morning, was three biscuits each. The Arabs are wonderfully abstemious, and are all the better for it.

"After I had returned to my tent I had a visit from Prince Halim Pasha, the Viceroy's brother, who had planted his tent about a league from our encampment. He informed me that the Bedouins whom he had sent out as scouts had told him that there were some herds of gazelles feeding at about two hours' march, and that when we started again he would get up a hunt.

"After breakfasting with the Viceroy and his brother, I have my horse saddled and take a ride round the outskirts of the camp. In one direction the desert extends beyond the horizon, while in the other it is bounded by Lake Mareotis, and farther on by the sea. Just under my horse's hoofs there jumps up a large jackal, which had, no doubt, taken up its quarters there so as to come and prowl about the camp at night, and after galloping in pursuit of it for ten minutes, and nearly touching it with the butt-end of my whip, it disappears among the brushwood.

"On my return I find the Viceroy sitting in front of his tent, and go with him to watch some shell practice with a mortar, but none of the gunners succeed in hitting the target five hundred yards off. The Viceroy, after listening to the band, which played the *Marseillaise*, among other tunes, had his dinner served in my tent."

"*November* 15, 1854.

"At five I was only partly dressed. Could any one have seen me outside my tent, with my red dressing-gown, like that of a Mecca cherif, performing my ablutions with my sleeves tucked up, they would have taken me for a true believer; and in the days of the Inquisition, as you know, one of the most serious offences was that of having washed one's arms up to the elbow. The camp gradually becomes more ani-

mated, and the chilly air announces the rising of the
sun, so I put on some warmer clothing and return
to my observations. A few rays of the sun begin to
illuminate the horizon, when suddenly there appears
in the west, where the sky is cloudy, a very brilliant
rainbow, running from east to west. I confess that
my heart beat violently, and that I was obliged to put
a rein upon my imagination, which was tempted to
see in this sign of alliance spoken of in the Scriptures,
the presage of the true union between the western
and the eastern world, and the dawning of the day for
the success of my project.

" The Viceroy's presence served to draw me out of
my reverie, as he came to wish me the ' top of the
morning,' and to ask me to take him round the country
which I had ridden over the day before. Preceded
by two lancers, and followed by the staff, we reached
an eminence where the ground was strewn with stones
which had formed part of some ancient building. The
Viceroy deemed this a very suitable place to prepare
for the morrow's start, so he sent an aide-de-camp to
have his tent and carriage brought up, the latter being
a sort of omnibus drawn by six mules and fitted up
as a bedroom. We rest under the shade of the carriage, while the chasseurs build up a circular parapet
formed of stones which they had picked up, and in
this parapet they make an embrasure into which a
gun is placed to salute the troops from Alexandria
which are just coming in sight. When I leave the

Viceroy to go and get my breakfast, in order to show
him how well my horse can jump, I put him over the
parapet and gallop off to my tent.   You will see that
this foolhardy act was one of the reasons which induced the Viceroy's *entourage* to support my scheme,
the generals who came to breakfast with me, and
who had seen the feat, telling me as much.

" I thought that the Viceroy had been sufficiently
prepared by my previous conversations to admit how
desirable it would be for a Government to have important public works of unquestionable utility executed by a financial company, and guided by the
happy presentiment of the rainbow I hoped that the
day would not close without a decision having been
come to with regard to the Suez Canal.

" At five o'clock I again mounted my horse and
came up to the Viceroy's tent by way of the parapet
He was very bright and good tempered, and taking me
by the hand, he led me to a divan and made me sit
by his side.   We were alone, and through the opening
of the tent I could see the setting of the sun which,
at its rising that morning, had so stirred my imagination.   I felt inwardly calm and assured at the moment
of entering upon a question which was to be decisive
of my future.   I had clearly before me my studies and
conclusions with regard to the canal, and the execution
of the work seemed so easy of realisation that I felt
little doubt as to being able to convince the Prince of
this.   I set out my project, without entering into

details, dwelling upon the principal facts and arguments set out in my memorandum, which I had by heart. Mohammed Said listened with evident interest to what I had to say, and I begged him if there were any points which did not seem clear to him to mention them to me. He, with considerable intelligence, raised a few objections, with respect to which I was able to satisfy him, as he at last said to me: 'I am convinced; I accept your plan; we will concern ourselves during the rest of our expedition as to the means of carrying it out. You may regard the matter as settled, and trust to me.' Thereupon he summoned his generals, bade them seat themselves upon some folding chairs which were just in front of the divan, and repeated the conversation we had had together, asking them to give their opinions as to the proposals of his 'friend,' as he was pleased to call me to these improvised advisers, better suited to give an opinion as to a cavalry manœuvre than a gigantic enterprise, the significance of which they were incapable of understanding. They stared at me and looked as if they thought that their master's friend, whom they had just seen put his horse over a wall, could not be otherwise than right, they raised their hands to their heads as their master spoke in sign of assent.

"The dinner was brought in upon a salver, and just as we had all been of one assent, so we all dipped our spoons into one and the same tureen, which contained some excellent soup. Such is the faithful and true narrative of the most important negotiation I ever undertook, or am likely to undertake.

" At about eight o'clock I took leave of the Viceroy, who told me that we were to start on the morrow. Upon regaining my encampment, Zulfikar Pasha guessed what had happened, and was not less pleased than myself. A friend of the Viceroy from childhood and his most intimate confidant, he had been of very material assistance to me in bringing about this happy result. I was not inclined to go to sleep, so I jotted down my travelling notes and gave a final polish to the memorandum which the Viceroy had asked me to draw up, and which had been ready for the last two years. It was as follows:—

" 'CAMP OF MAREA, *November* 15, 1854.

" ' The joining together of the Mediterranean and the Red Sea by a navigable canal is an enterprise the utility of which has attracted the attention of all the great men who have reigned or been for a time in Egypt: Sesostris, Alexander, Cæsar, the Arab conqueror Amrou, Napoleon I., and Mohammed Ali. A canal communicating by way of the Nile with the two seas existed in ancient times—but for how long we know not—under the old Egyptian dynasties; during a second period of 445 years, from the first successors of Alexander and the Roman conquest until the fourth century before the Hegira, and finally for a period of 130 years after the Arab conquest.

" ' Napoleon, directly he arrived in Egypt, appointed a commission of engineers to ascertain whether it would not be possible to re-establish and to perfect this mode of communication. The question was decided in the affirmative, and when the learned M. Lepère handed him the report, he said : " The work is great, and though I shall not now be able to accomplish it, the Turkish Government will some day, perhaps, reap the glory of carrying it out.

" ' The time has come to realise Napoleon's prediction. The piercing of the Isthmus of Suez is certainly an enterprise destined to contribute more than any other to the maintenance of the Ottoman Empire, and to demonstrate to those who announced its approaching ruin that it still has a fruitful existence before it, and that it is capable of adding a brilliant page the more to the history of human civilisation.

" ' Why is it that the Governments and people of the West are at the present moment coalesced to maintain the Sultan in the possession of Constantinople ? and why does the Power which has threatened to deprive him of it meet with such opposition if it be not that the passage from the Mediterranean to the Black Sea is of such importance that the European Power which became mistress of it would overrule all the rest ?

" ' If a similar and still more important position were established in another part of the Ottoman Empire, and if the commerce of the world were made to pass through Egypt, the position of the Empire would be doubly strengthened, for the great European Powers, in order to prevent any one of them ever seizing it, would regard it as a vital necessity to guarantee the neutrality of the canal.

" ' M. Lepère, half a century ago, estimated that the re-opening of the old roundabout canal would take ten thousand workmen four years, and would cost from twelve to sixteen hundred thousand pounds. He also considered practicable the piercing of the Isthmus of Suez at Pelusium.

" ' M. Paulin-Talabot, one of the three celebrated engineers selected ten years ago, together with Messrs. Stephenson and Negretti, by a company which had been formed to prosecute this enterprise, was in favour of the indirect route from Alexandria to Suez, making use of the dam for crossing the Nile. He estimated the total cost at £5,200,000 for the canal, and £800,000 for the port and harbour of Suez.

" ' Linant-Bey, who for the last thirty years has been conducting with much ability the canalising works in Egypt, who has made this question of the junction of the two seas his constant study, and whose opinion, formed from practical experience, deserves the utmost consideration, had proposed to pierce the isthmus almost in a straight line at its narrowest point, forming a large inland port in the basin of Lake Timsah, and making the passes of Pelusium and Suez available to ships of the largest tonnage from the Mediterranean and the Red Sea.

" ' The general of engineers, Gallice-Bey, the director and designer of the fortifications of Alexandria, had, upon our initiative, submitted to Mohammed Ali a scheme very closely resembling that of Linant-Bey. Mougel-Bey, director of the dam works upon the Nile, chief engineer of roads and bridges, had also spoken to Mohammed Ali of the possibility and utility of the piercing of the Isthmus of Suez ; and in 1840, at the request of Count Walewski, then upon a mission in Egypt, he was instructed to take preliminary steps in Europe, which the course of political events unfortunately cut short. Careful examination will show which of the routes is the best, but as the enterprise has been proved to be practicable, all that remains is to select the most feasible project.

" 'No operation, however difficult, is now regarded by modern art as impossible. Its feasibility is not doubted; it is merely a question of money, which the spirit of enterprise and co-operation will soon solve, if the profits which are to result from it are in proportion to the cost. It is easy to show that the cost of the Suez Canal, taking the highest estimate, is not out of proportion with the usefulness and the profits of this great work, which would abridge by more than half the distance between the principal countries of Europe, America, and the Indies.

" ' Mohammed Said has not been slow to see that there was no work which, as regards the grandeur and utility of its results, could compare with this. What a glorious record for his reign, what an inexhaustible source of wealth for Egypt it will be ! The names of the Egyptian sovereigns who erected the pyramids, those monuments of human pride, remain unknown. The name of the Prince who opens the great maritime canal will be blessed from century to century, down to the most distant posterity.

" ' The pilgrimage to Mecca secured for all time and made easy for the Mohammedans ; an immense impulse given to steam navigation and long voyages; the countries along the Red Sea and the Persian Gulf, the eastern coast of Africa, India, the kingdom of Siam, Cochin China, Japan, the vast Chinese Empire, the Philippine Islands, Australia, and that vast archipelago towards which the emigration of ancient Europe is tending, all brought three thousand leagues nearer to the basin of the Mediterranean, as well as to the north of Europe and America : such are the sudden and immediate results of piercing the Isthmus of Suez.

" ' It has been calculated that the navigation of Europe and America, by the Cape of Good Hope and Cape Horn, is equivalent to an annual movement of six million tons, and that if only half of this was conveyed by way of the Gulf of Arabia, the trade of the world would realise a profit of six millions sterling per annum.

" ' It is beyond doubt that the Suez Canal will lead to a great increase of tonnage, but reckoning only three million tons, a tax of ten francs a ton, which might be reduced as the navigation increased, would yield annually £1,200,000.

" 'In terminating this memorandum, I think it right to call your Highness's attention to the preparations which are being made for opening up communications between the Atlantic and Pacific Oceans, and to the effects upon the world's trade, and ultimately upon the future of Turkey, by the opening of these new routes if the isthmus which separates the Mediterranean and the Red Sea were to remain much longer closed to trade and navigation.

" 'Does not this show that the time for treating that question has arrived? May we not conclude that this great work, far more important for the future of the world, is henceforth secure from all serious opposition, and that the efforts made to realise the project will be sustained by the universal sympathy and active assistance of the enlightened men of all countries?

" ' F. DE LESSEPS.'

"The ground at Marea, where we are encamped (*Gheil* in Arabic) has still some traces of antiquity. I remarked the shafts of several columns and an immense cistern, half destroyed, with ten or twelve pointed arches, while all over the surrounding hills are to be seen stone which has been used for building. There is an abundance of good water, of which Ibrahim, a character of whom it is worth while to give a sketch, had brought us some from the cistern. Ibrahim

is a very good specimen of the cunning and greedy
Arab, such as Europeans are familiar with. He had
met Clot-Bey in the streets of Alexandria, had made
out that he had formerly been under his treatment,
and had declared his intention of serving him in
future. In this way he had joined the camp, with
the new master to whom he had, so to speak, attached
himself by force. As soon as he saw me, he found
that I was the most influential person in the camp, so
he transferred his attentions to me, telling me that 'I
was in his eye' and that he had become attached to
me, and would not leave me while I was in Egypt.
This sudden change of views, which I communicated
to my companion, was not calculated to give me a
high opinion of Ibrahim's morality; but he was so
careful in anticipating all our wants, from morning to
night, and so intelligent in his service, that we left
him to look after all the details of our encampment, to
strike the tent and have it loaded with the rest of the
luggage on the camels, and to have it got ready at the
next halting-place, where he was always waiting for us
with some fresh water and a cup of coffee.''

" *November* 16, 1854.

" Being the first to get up, I take advantage of this
to write and communicate the good news to France,
Zulfikar sending off the letters by a messenger on a
dromedary to Alexandria. After breakfasting with
the Viceroy, a signal gun announces that the camp is

about to be raised, and in the twinkling of an eye thousands of tents are struck and placed upon the camels' backs. This large caravan defiles past us, leaving Lake Mareotis in its rear and taking the desert road, looking like a large piece of ribbon being unwound. The infantry regiments are formed in three columns, with sharpshooters on either side, followed by the artillery and cavalry. The Viceroy follows on horseback, with me to his right, and Selim Pasha, the general of cavalry, to his left. Selim is one of the former students of the school of Giseh, whom I remember entering the service in 1833 under the French Colonel Varin. We gallop off from the hill on which we were to a plateau just opposite, and we watch the army defiling below us, the soldiers cheering and brandishing their muskets as they pass before the Viceroy, the cuirassiers, wearing the ancient Sarrazin helmets, which glittered in the sun, looking remarkably well. After the march past we took our places at the head of the army, preceded by a dozen Bedouin horsemen who acted as scouts. There had been no army on the march in this region since the expedition of General Bonaparte, whose brave troops underwent great hardships, where we were merely making a military promenade, with every conceivable comfort provided for us.

" Upon arriving at our halting-place for the night, the Viceroy sent to say that he was tired and was going to bed; but that he would send me the dinner

which, by the light of a dozen torches, a troop of some
five-and-twenty or thirty cooks and scullery lads were
getting ready. I went to have a look at this open-
air laboratory. Three rows of saucepans, placed in a
row over some trenches which had been dug in the
ground, were being heated by faggots placed in the
hollow of the ground. This is not an economical
mode of cooking, but it is a very expeditious way.
After dinner our tent is converted into a drawing-
room, for it is gradually becoming the rendezvous of
the staff, who come to hear the news, while Zulfikar
carries on his correspondence, opens the Viceroy's
letters, receives and despatches the messengers, and
gives the orders in the Prince's name."

                                    " *November* 17, 1854.

" At seven the Prince was up and out of his tent,
and upon my going out to him he tells me that he
has been disturbed by the trumpeters of the cavalry,
who are quite close to his tent, which he has placed
about 350 feet farther off, the intervening space being
intended for the erection of targets at which he means
his artillerymen and chasseurs to practise. The day
is spent by the troops in bathing and washing their
clothes in the canal; and, after a ride on my horse, I
come to where the Viceroy is making his sharpshooters aim at a target about 550 yards off. None
of them had as yet hit it, so, taking the carbine from
one of them, I showed them how to shoulder it and

how to fire. The officer asked me to try a shot, and I hit the bull's-eye. The Viceroy then sent for his own carbine, one of German make, and with that too I hit the mark; but I declined to go on again, so as not to endanger my reputation of being a good shot. After breakfast we form a circle round the tent, and the Sheik Masri, who had shown much devotion to the Viceroy when he was persecuted by Abbas Pasha, and whom he had subsequently attached to his household, related to us the war which had occurred six months before, between a tribe from Upper Egypt and that of the Ouled-Ali, to which he belonged. The Ouled-Ali tribe encamps in the deserts which extend from Lake Mareotis to the seashore as far as the frontiers of Tripoli, cultivating the land bounded by the last of the canals which separate the desert from the provinces of Lower Egypt. The Ouled-Alis, which, with a population of only 50,000, have 10,000 guns, expecting an attack from their enemies whom the policy of Abbas Pasha had raised up against them, had formed a corps composed of 6,000 men and a certain number of women whose mission it was to urge on the men to combat by their shouts and songs. In action they are mounted upon camels, and more exposed to danger than the men. The Ouled-Alis formed entrenchments with sandbags and fascines near the village of Hoche, which we shall pass to-morrow. Here they made a stand against their opponents, who lost two hundred men in the attack, while they had four women and three men killed. The Bedouins of Upper Egypt fled and did not return."

" *November* 18, 1854.

" We start two hours before daylight, the Viceroy having preceded us. We overtake him about ten o'clock at Hoche, where he was awaiting us in the tent of the governor of the province.   More than a hundred Bedouin chiefs of the Ouled-Ali tribe are assembled here; they are all men of high stature and appear very quick and intelligent.   The troops arrive and get under canvas, the heat being terrific, and shade most grateful.   The chief who was in command during the combat referred to above comes to pay us a visit, accompanied by his son, who is as tall as himself. Prince Halim Pasha joins us, and says that we have left the region where the gazelles were to be found considerably to the right, but I am not sorry for this, as I am anxious to be as much as possible alone with the Viceroy to talk over my plans.

" We start at three, preceded by a troop of Bed-ouins, who every now and again start off at full gallop, wheel round, and fire, this being what they call a *fan-tasia*.   Reaching Zaoui-el-Khamour at sunset, the Viceroy, whose hours for meals are very irregular, tells me not to wait dinner for him, and after having had it served in my tent I was just going off to sleep, about nine o'clock, when I heard the sound of female singing, mingled with the beating of tambourines and castanets. Paolini Bey came to fetch me on the part of the Viceroy, who had allowed a troop of almées (dancing girls) to come and perform. He gave me a place on his divan, the almées crouching in a circle upon the carpet. One of them was richly attired, and had, so the Viceroy informed me, more than £400 worth of embroidery and jewellery upon her. They recommenced singing, and every now and then the Kaouadji, or chief coffee man, gently struck the singers upon the cheek, as you might a child, and made them swallow sweets and syrups. After the singing was over two of the almées got up, and standing opposite to each other, like Spanish ballerinas, began to execute their dances. Two others followed, after which the whole troop filed past the Khedive, and respectfully kissing his feet, were dismissed."

" *November* 19, 1854.

" We start at seven and halt at nine. The Viceroy quits his horse for the carriage, so we go on ahead with Halim Pasha, reaching at noon a regular Egypt-ian village, called Yahoudié. We pass over a dyke and reach a small island, situated in the midst of a cultivated and partially irrigated plain, where we find a delightful shady spot, with sycamores, willows, and mulberry trees, forming a belt of verdure around a small lake. This oasis presents a charming contrast with the sandy hillocks which we have just come over, and which will prove very embarrassing to the artillery.

"At two o'clock the Viceroy arrives, followed by his battalion of chasseurs, and upon my going to see him I hear that he has given orders for ten steamers to be collected at Neguileh, upon the Nile, where we shall be to-morrow, and he tells me that when on the steamer he shall want me to read him my memorandum upon the Suez Canal.

"The Viceroy being about to go to bed, on account of our early start in the morning, I left him and dined in my tent with Halim and the generals."

"*November* 20, 1854.

"The Viceroy did not get up so early as he had intended, for he had passed a bad night. News having been brought to him that the artillery could not get through the sand, and that several horses were already dead of fatigue, he sent reinforcements, and by dint of hard work the guns were got through.

"At eight preparations were made for a start, and while our tent was being struck an eagle came and hovered over us. Zulfikar handed me his gun, and aiming at it, I brought the bird down dead at my feet. If I mention this incident, so insignificant in itself, it is because it is destined to have an influence upon public opinion in Egypt as regards the success of my enterprise. We mount our horses and accompany the Viceroy to a village, where we alight under the shade of two sycamores, with a delicious carpet of greenery around us, for it is at this season that the wheat is green. I leave the Viceroy there and ride on to overtake Halim. Soon after reaching the magnificent river, which, we are told, has such an irresistible attraction for the stranger who has once drunk its waters, a boat takes us out to the yacht which the Viceroy's predecessor, Abbas Pasha, had built in England at a cost of £100,000. It is quite beyond me to describe the luxurious character of the fittings, the painting, and the furnishing of this vessel, with its doors in oak and citron wood, its locks and fastenings in solid silver, its medallions representing rivers and animals painted by distinguished artists, its staircases with silver balustrades, its divans lined with cloth of gold, its dining-room forty feet long, and its bedrooms like those of a palace. The Viceroy comes in soon after, and after again showing me this floating palace, says that of course he should never have committed such an act of folly as to build such a boat, but that as she is in existence he makes use of her. He places at my service during the two or three days we spend on the Nile to wait for the troops and despatch them to Cairo his ancient old steamer the *El Ferusi* (the *Turquoise*), assigning another boat to Halim Pasha. My quarters on the *Turquoise* consists of a saloon forty feet long, with a large divan decorated with handsome Lyons silk brocaded in gold, of a bedroom, a dressing-room, and a bath-room in white marble. Clot Bey,

Hassan Pasha, and two generals, occupy other rooms, and we dine in the saloon."

" ON THE NILE, *November* 21, 1854.

" I pay a visit to the Viceroy in the morning and read him my memorandum, striking out one or two passages of which he did not approve. I also read him the draft firman of concession, which he approves in its entirety.

" After a talk with Halim Pasha, he promises me to do what he can to effect the reconciliation of Achmet Pasha, the heir to the Viceroy, and the latter. We next go to see Mahommed Said, where we breakfasted. We arranged for a hunting party in a village close to Néguileh, and the Viceroy had the steam got up on his boat. Upon reaching the island, where we had been told that we should find wild boars, we saw plenty of traces of them, but they probably only came there of a night, so we returned to our previous anchorage. A steamer from Cairo, with passengers from Alexandria, landed Moustafa Pasha, a brother of Achmet Pasha, and a nephew of the Viceroy, who asked him to join us in our expedition. He also told him about my scheme and advised him to read my memorandum."

" *November* 22, 1854.

" Had a conversation with Moustafa about the canal. He is very intelligent and well-informed, and speaks French like a Parisian. We breakfast on the Viceroy's yacht. His Highness informs me that the *Turquoise* will that evening take on board troops and start in the night for Cairo with all the other vessels except his own, which will leave in the morning. So I have my things moved on board."

"*November* 23, 1854.

"When I go early on deck, Moustafa begs me to read him my memorandum. He seems very satisfied with it, shows much enthusiasm for the undertaking, and says that he will put money into it. The Viceroy, who has joined us, himself opens the subject of the canal, asking me what engineer is to make the preliminary investigations upon the spot. I tell him Linant Bey, to whom Mougel Bey might also be adjoined, and that their report can then be considered by the English, French, and German engineers, whose researches will be submitted to the commission over which I shall preside, and which will decide as to the best route to follow.

"At night time we stop for an hour at the dams, which we see by torchlight, reaching Boulac at eleven and spending the night there."

"CAIRO, *November* 24, 1854.

"I rise at six and find the Viceroy has already gone off incognito to the citadel. He had told Zulfikar that I was to wait for a carriage which would come to take

me to the Palace of the Muçafirs (strangers), near the
mosque of Setti-Zeneb (St. Zenobia), where apartments had been retained for me. At seven a large
barouche with four horses and two chiaous (officers
of the Viceroy's household) carrying their silver-
headed canes drove up, and I made it halt near the
Square of Esbekié, at the house of Linant Bey, who
threw himself into my arms when I told him that his
dream of the piercing of the isthmus was about to
become a reality. I went up also to see Madame
Linant, whose marriage I had when French consul
at Cairo celebrated, but whom I had not set eyes on
since.

" Lubbert Bey, Secretary of the Ministry of Foreign
Affairs, who lived in the neighbourhood, hearing of
my arrival, came to see me. When I first visited
Alexandria as student-consul in 1832, Lubbert was
the friend and guest of my dearly beloved superior,
M. Mimaut, one of the most distinguished diplomatists ever in the service of France. I shall never
forget how M. Mimaut, with the great work on the
Egyptian expedition in his hand, gave me the first
notion of the canal between the two seas, a subject
of which up to that date I had not the faintest knowledge.

" I then go on to take possession of my new abode,
and make a grand entrance between two rows of
mamelukes and servants. The nazir (steward) of
the palace, a worthy effendi with a grey beard,

reminding one of Francis I., rushes to the door of the carriage and holds me by the arm as I get out, and so escorts me to the apartments, followed by the chiaous and other servants. The Palace of the Muçafirs was the residence of the Egyptian Institute at the time of the French expedition, and it was there that the commission of savants who had been ordered to report upon the canal used to meet. It was a singular coincidence that after the lapse of half a century these same walls should witness the realisation of a work which had been thought out within them at the bidding of the greatest man of his century.

" I am informed that I have twenty horses at my disposal, ten for harness and ten for riding, a state coach richly gilt, a barouche, a landau, and a 'my lord' ! Breakfast was laid for twelve covers.

"The Viceroy had advised me to lose no time in going to see Mr. Bruce, the agent and consul-general of England, to tell him of his Highness's intention to make the canal, and to communicate to him the documents relating to it.

" I have a conference of two hours with him, and he told me that though he could not speak for his Government, to whom he would report my visit, he did not hesitate to give me his personal opinion, which was that so long as there was no intervention on behalf of any foreign Power in the affair, and that the work was carried out by means of capital freely subscribed to an enterprise sanctioned by the governor of the country, he could not foresee that any difficulty would be raised by England.   I replied that I was of the same opinion, and that the question, formerly so serious, of the opening of the Suez Canal being now extricated from the political difficulties which had obscured it, became a mere matter of practical possibility and finding the money.   As regards the practicability of the scheme, men of science had already pronounced in its favour, as others would do; while as regards money, it is certain not to run short for a work which will not only enrich the trade of the world, but will, according to the most modest estimate, be a profitable speculation for shareholders.   It is agreed that I shall send a letter to Mr. Bruce, enclosing a copy of my memorandum and of the firman, and the Viceroy expresses himself as satisfied with the course of this interview.

" Kœnig Bey, the former tutor to the Viceroy, who is now his private secretary, is ordered to translate the documents relating to the canal into Turkish."

" CAIRO, *November* 25, 1854.

" 'The Viceroy had asked me, without giving any reason, to go to the citadel at 9 A.M., and upon entering the grand divan I find the Viceroy seated at the same spot where his aged father, Mehemet Ali, had often received me, and where he once told me the story of the massacre of the Mamelukes.   All the functionaries came to congratulate the Viceroy upon his safe arrival in his capital, and no sooner had they taken their seats on the divan than he publicly declared that he had resolved to open up the Isthmus of Suez by means of a maritime canal, and to entrust me with the formation of a company composed of capitalists of all nations to which he would cede the right to execute and work this enterprise. Then, speaking to me, he said, 'Is this not so?' I then spoke a few words, taking care to let the spontaneity and merit of the decision remain with him to avoid ruffling the susceptibilities of foreigners.

" The Consul-General of England was somewhat ill at ease, but the Consul-General of the United States, to whom the Viceroy had said, 'Well, M. de Leon, we are going to start an opposition to the Isthmus of Panama, and we shall be done before you,' had boldly spoken out and replied in what I could not but regard as a favourable sense.

"After the consuls had withdrawn, I told the Viceroy of the coincidence of my being lodged in the residence of the ancient Egyptian Institute, and it struck him as being so strange that he sent for several of his intimate friends to tell them of it. He was very satisfied at having made this declaration to the consuls. I told him that I should never have dared to advise him to do it, but that I thought he had taken the best course for cutting short a great many objections and difficulties by letting public opinion know of a project the general utility of which is incontestable.   His reply was : 'I must admit that I had not thought of this ; it was an act of sudden inspiration ; you know that I am not inclined to follow ordinary rules, and that I do not like to do as other people do.'

" While I am conversing with the Viceroy, Solimon Pasha comes to see him on military matters, and I drove off in my state coach drawn by four white horses. The negro coachman drives very well, and goes at full trot or in a gallop through the narrow streets and bazaars of Cairo, though I must add that the footmen distribute, in spite of my admonitions, blows with their staves right and left to keep off the persons on foot, who stand close up against the walls and shops.   These poor fellows do not complain, but on the contrary exclaim in a tone of admiration, 'Ah; there is a grand seigneur going by.   Mashallah (Glory be to God)' !

" Such is the East, such it has been from all time, and so it is described to us in the Bible, where we read that after Joshua had massacred the inhabitants of Jericho, even to the women and the asses, 'so was the might of the Lord made manifest.'

" In the course of the day I go to see the three sons of Ibrahim Pasha, the eldest of whom, Achmet Pasha, is a well-educated man who had distinguished himself at the French Polytechnic School.   He is very well qualified, like his father, to administer his vast pro-perties, and he argues very well in French upon all topics.   He had been to see the Viceroy on the morning of his arrival, and had been very well received. He knew that I had helped to effect the reconciliation, and he was grateful accordingly.

"I have already alluded to Ibrahim Pasha's third son, Moustafa, and as to the second, Ismael, I like him very much, and am delighted at his reception of me. He has handsome and distinguished features, and shows all the blood of Mehemet Ali. When he comes to think less of his pleasures, he will, I think, made his mark in the world. Although only five-and-twenty, he is already the father of twelve children. He inherited from his father the finest palace in Cairo, upon the banks of the Nile, and he has spent there more than £40,000 upon furniture which he has had sent from France. He showed me through his vast and splendid apartments on the ground floor, and part of those on the first floor, the rest being reserved for the harem. Passing through a large saloon, I could see the hangings heaving to and fro behind which the eunuchs were moving about. The banisters of the staircase are of carved rosewood, encrusted with silver, the balustrades being of Baccarat crystal.

"From thence I went to the house of Halim Pasha, who inhabits one of Mehemet Ali's residences. There is an avenue leading up to it, a league in length, of large sycamores, which I remember as very fine trees when I first saw them, but which now form a dense roof of greenery. This avenue was planted by the French army in 1800. Halim received me very graciously, and expressed himself as delighted at his brother's declaration with respect to the Suez Canal. He has all the vivacity and manners of a southern Frenchman, with a very pure Parisian accent.

" I also paid a visit to M. Huber, the Austrian agent and consul. He spoke to me of the interest his Government took in the opening of the Suez Canal, and said that he had been instructed to support the project very heartily when it came to be discussed. He afterwards came to dine with me, when he met Mr. Bruce, Baron von Pentz, consul-general of Prussia, Count d'Escayrac, a French traveller, Linant Bey, and others.

" Clot Bey has become my guest, the Viceroy having told Zulfikar to ask him to come and stay here. He has introduced to me M. Reynier, a young poet, who is tutor to his children, and whom he has brought from Marseilles. M. Reynier is two-and-twenty, and he has very charming features, with an open and candid air which has much prepossessed me in his favour. His father is librarian to the town of Marseilles, and he has very kindly placed himself at my disposal as secretary. After having made a few copies of my memorandum and firman he was able to write them off by heart."

" November 26, 1854.

" I receive a visit from Talat Bey, the Viceroy's first secretary for Turkish affairs, and Kœnig Bey, who occupies the same post for European affairs, came with him as interpreter.

"At ten o'clock I go up to the citadel, where the Viceroy keeps me to breakfast, the conversation turning upon what he calls 'my affair.' It is arranged that Mougel Bey shall take part in the exploring which we propose to make with Linant. At first he made some objections as to the difficulty of keeping two engineers of one mind, but at last he agreed to my proposal, by which I set considerable store. Upon my return to Setti-Zeneb I receive a visit from Achmet Pasha, whom I like more each time I see him. Among other visitors was Arnaud Bey, who had just travelled 1,200 leagues up the Nile, or 300 further than anyone else. He gives us some very interesting details of the expedition, which had been organised by Mehemet Ali, who had given him a corps of 800 men. He could not go any further owing to the mutinous attitude of some of the officers. All the expenses had been defrayed by elephants' tusks, which they brought back by water. Arnaud Bey spoke in high terms of the treatment which he had met with from the negro populations he had traversed, and which had never before seen boats with sails. None of his men had been killed by the natives. Remounting towards the Equator, to which he got within two degrees, the natives told him that navigation was possible up to the fourth degree beyond the Equator, that is to say for another 150 leagues. In the higher regions on both banks of the Nile there are immense forests full of elephants, lions, and animals of all kinds. His troops had sometimes fired into herds of hundreds of elephants, which went tranquilly on their way without looking round, and took no more notice of the firing than if they had been pelted with sweatmeats. Upon one occasion an elephant was surrounded, whereupon he rushed at one of the men, seized him with his trunk, and hurled him into the air; at last, by dint of firing at him at close quarters, the animal was killed.

To Mr. Bruce, Agent and Consul-General of Her Britannic Majesty in Egypt.

"*November* 27, 1854.

" I have already had the honour of speaking to you about the Viceroy of Egypt's project for piercing the Isthmus of Suez. His Highness, who intends to make over to me his powers for the constitution of a Universal Company, to which would be conceded the making and working of the new route, has requested me to communicate to you a copy of the memorandum which he has asked me to draw up on this subject, in which he wishes to meet the wishes of England as well as of other nations. Anything which contributes to the extension of trade, industry, and navigation must be specially advantageous to England, considering that she takes rank before all other powers in the importance of her navy, her manufacturing products, and her commercial relations. Only the unfortunate prejudices which, owing to political differences, have so long divided France and England could have accredited the belief that the opening of the Suez Canal, a work of civilisation and progress, would be detrimental to British interests. The frank and sincere alliance of the two nations which are at the head of civilisation, an alliance which has already proved the possibility of solutions heretofore regarded as impossible, will facilitate, among other beneficial results, an impartial consideration of this vast question of the Suez Canal, will enable us to form a true estimate of its influence upon the prosperity of all nations, and will prove that it is a heresy to believe that an enterprise destined to shorten by a full half the distance between the east and the west is not good for England, the mistress of Gibraltar, Malta, the Ionian Islands,* Aden, and important establishments upon the east coast of Africa, India, Singapore, and Australia.

England, therefore, as much as and even more than France, must be in favour of piercing this narrow strip of land only forty leagues in breadth, which no one who gives a thought to questions of civilisation and progress can see on the map without earnestly wishing to wipe out the only obstacle in the way of the main route for the trade of the world.

"The communication of my memorandum, and of the powers which the Viceroy proposes to confer on

* Note of the Translator.—They have since been ceded to Greece.

me, obviate the necessity of my entering more into
detail about an enterprise in which, as you will observe, there is no question of any special privileges
for one state more than another, all that is intended
being to constitute an independent company, which
the shareholders of all nationalities will be able to join
upon equal terms."

To Mr. Richard Cobden, M.P., London.

"CAIRO, *December* 3, 1854.

"As a friend of peace and of the Anglo-French
alliance I send you a piece of news which will contribute to realise the saying 'aperire terram gentibus.'
I refer to the Viceroy's concession of powers for making
a canal through the Isthmus of Suez. Some persons
assert that the project will excite hostility in England.
I cannot believe it. Your statesmen are too enlightened for me to admit such an idea. What! England
has herself one-half of the general trade with the
Indies and China; she possesses an immense empire
in Asia; she can reduce by a third the costs of her
trade and reduce by one-half the distance; and she
will refuse to do so, simply in order that the nations
bordering on the Mediterranean may not benefit by
their geographical situation to do a little more trade
in Eastern waters than they do at present! She would
deprive herself of the advantages to be derived materially and politically from this new mode of communication, merely because others are more favourably

placed than herself, just as if the geographical situation was everything, and as if, taking everything into
account, England had not more to gain from this work
than all the Powers put together. Then, again, we
are told that England apprehends that the diminution
by more than a third in the voyage to India would
lead to a reduction in the number of merchantmen.
The experience of railways has surely proved to an
extent exceeding the boldest estimates that a shorten-
ing in the distance and an abbreviation in the length of
a journey increases to an extent exceeding all calculation the business relations and traffic. It is wonderful
that those who raise this objection do not advise the
English Government to send ships to India by way of
Cape Horn, as that would entail a still further increase
in the number of ships, the distance being so much
longer. If by any possible chance the difficulties
with which we already are threatened should arise, I
hope that the public spirit which is so powerful in
England will soon override interested opposition and
antiquated objections.

" Let me hope also that, should the occasion require
it, I may count upon your support."

" *December* 17, 1854.

" I pay a visit to Father Leonardo, superior of the
Franciscan Monastery, who receives me in the room
occupied by Murat during the French expedition.
While making a cruise on the Nile Kœnig Bey shows

me a letter he had written to Mr. Bruce, asking him
to come on board at eleven and bring Mr. Murray
with him. The Viceroy had very reluctantly agreed
to give Mr. Murray an audience, as he accused him of
having instigated Abbas Pasha to persecute him in
bygone years, but I advised him to do so rather than
offend the English agent. The audience passed off
very smoothly, and after the departure of Mr. Murray,
with whom he shook hands, the Viceroy said laughingly
to me, 'I shall not give my hand to a friend to-day,
having done so to an enemy. I showed myself a good
diplomatist, did I not, and said many things which I
did not think?'

"The Viceroy has heard with much satisfaction
from M. Sabatier, the French consul, that the Emperor has sent him an autograph letter, accompanied
with the Grand Cordon of the Legion of Honour."

" *December* 18, 1854.

"I pay a visit to the Viceroy, who is returning to
Tourah. He has heard that I had told Linant to draw
money from my bankers to pay for the utensils and
provisions required for our journey through the isthmus. He said that he could not allow this, and he
very quietly chided me for having supposed that he
would permit me to incur the least expense while I
was his guest.

"At Setti-Zeneb this evening I gave a dinner of
thirty covers, in the Viceroy's name, in honour of Mr.

Murray, who was invited, together with all the consuls-
general, his late colleagues, several Englishmen and
Egyptian functionaries."

" December 19, 1854.

"Kœnig Bey fetches me this morning to go and see
the Viceroy at Tourah, where I spend part of the day
in discussing our journey to the isthmus, which is
definitely fixed for the 23rd."

" December 22, 1854.

"The reception of the Grand Cordon of the Legion
of Honour took place to-day at the citadel, the presentation being made by M. Sabatier, to whose con-
gratulations the Viceroy replied in a neat little speech
delivered in very good French."

" December 23, 1854.

"We start at nine, M. and Madame Sabatier and
a few friends accompanying me as far as Suez. We
take with us two mounted couriers, one for the relays
at the different stations, the other to accompany our
carriages. There is a good macadam road from Cairo
to Suez, and fifteen relays, or stations, and after dining
and sleeping at the eighth we reach Suez at noon the
next day, having done our thirty-three leagues in
the desert as comfortably as if we had travelled from
Paris to Orleans."

" SUEZ, *December* 25, 1854.

"The rising sun lights up my room, and, opening
my window, I gaze in mute contemplation upon the
Red Sea, whose rising waters lap the walls of the

Hôtel des Indes, with Mount Attaka to the right, while to the left I can just make out the beginning of the mountain chain which culminates in Mount Sinai. This part of the coast has a rosy tint which is reflected in the waters, whence I suppose comes the name of ' Red Sea.' People are beginning to move about on the quay, and boats, the oars of which are long poles with a round paddle at the end, accost the vessels which have just arrived or are leaving for Jeddah. These boats, with no decks, but with an elevated poop and painted prow, are not unlike the Chinese junks. The dresses of the natives and the foreigners, as well as the furniture of the houses, give the traveller a forecast of what Arabia, India, and China are like. I notice that the inhabitants are more deliberate in their movements than in the rest of Egypt.

" Suez is, moreover, an isolated point, surrounded by deserts ; its population, numbering from three to four thousand, is a very miserable one, having only brackish water to drink. Our canal will bring it water and activity, which it lacks.

" Going up to the terraced roof of the hotel I am able to obtain a complete topographical notion of the surrounding country, and I am anxious to see all this for myself, as when I have taken in a thing myself I shall be able to make it comprehensible to those who are not engineers. Linant and Mougel ask me to be sure and always give them my opinion. I had been told that perhaps they would not always agree, but in

an affair of this importance it is desirable to have two
independent opinions, even if they do not concur.
Linant knows the topography of the whole country;
he has made a map of it, and has studied the geology
of it on the ground. The whole canalising system of
Egypt is familiar to him, while Mougel, on the other
hand, has carried out important hydraulic works in
Egypt; and though no one can pronounce so well as
Linant upon the direction which the canal shall follow,
the opinion of Mougel will be preponderant upon the
question, not yet settled, as to the point of entry, both
on the Mediterranean and the Red Sea.

"My companions have not yet made their appearance, so I go and wake them up and propose that in
the course of the day we should make an excursion
into the desert as far as the beginning of the canal
which had been made by the ancient kings.

"We start after breakfast, some on horseback and
others in carriages, escorted by fifteen bashi-bazouks,
and when we reach the spot we find that the banks of
the canal are still distinctly visible. Measuring the
bed, we find that it has just the breadth of 90 cubits
spoken of by Herodotus. Upon our return we rest in
one of the tents prepared for our journey. Linant
has some excellent coffee served us by his maître
d'hôtel, the aged negro Abdallah. My friend Ibrahim
has been relegated to obscurity, for his head had been
turned by the exalted position to which he had
attained. In order, I suppose, that his appearance

should be in keeping with it, he had bought a fine
gilt sword, a wand of office, patent leather shoes, and
a very gaudy sash, but in order to pay for them he
had taken the money he wanted out of my pockets
and from other people's. So, without more ado, he
had been turned out of the place and told to show his
face no more."

"*December* 26, 1854.

"The journey by land from Suez to the Fountains
of Moses takes more than two hours, so we embark
upon a government steamer, which lands us there in
half the time.

"M. Costa, the owner of the principal spring,
accompanies us and tells us that he has had a breakfast prepared for us at which a sheep will be served
whole. His wife and sister-in-law, attired in the
richest of Oriental costumes, the eyelids and lashes
painted, accompany Madame Sabatier. We had
scarcely emerged from the narrow pass into the gulf
before a very strong wind got up. The captain and pilot
declare that it would be impossible to land, which we
can quite believe when we see how the waves dash
upon the shore. The road is covered with eddies of
sand which would have made the drive most unpleasant for the ladies, so we have to give up the
excursion, and M. Sabatier prepares to return to
Cairo, taking with him all the persons who had
accompanied us to Suez."

" December 27, 1854.

" We spend the day in a careful examination of the port of Suez and the mouth of the canal. When the tide is low, we go out to the islets and to what look like beds of rocks, but which we find to be remains of ancient masonry. We break off fragments at the spot which probably formed the floating-dock of ancient Chlysma. I intend to have it analysed by M. Le Play, sending him at the same time specimens of the stone and materials which may be taken from the neighbouring mountains. In a small island near the port the East India Company has formed a cemetery, but it has been found necessary to surround it with a wall, as the Arab women who had no children used to go and steal the bones of the Christians, which, worn as amulets, are considered to ensure child-bearing.

" We dined with Mr. West, the English consul, and our dinner is composed of mutton from Calcutta, potatoes from Bombay, green peas from England, poultry from Egypt, water from the Ganges, wine from France, coffee from Moka, and tea from China."

" December 28, 1854.

" We mount our horses at eight and ride off to the principal gorge of Mount Attaka, where St. Anthony is said to have lived in a grotto, which is no longer to be seen. When one leaves Suez, it seems as if the mountain is within half an hour's ride, but it takes us at least three hours.

" We follow the bed of a torrent which leads us to the gorge where we intend to make our collection of minerals, and we get together specimens of marble, calcareous marl, clay, &c., which we carry back with us to Suez, where M. Costa takes us to see the house in which Bonaparte lodged. The present owner, born in the same year, has commemorated the visit by decorating his salon with drawings of all the great victories gained during the First Empire."

" December 29, 1854.

" Early in the morning we go by steamer to the extremity of the Bay of Suez, to the shore where terminates the valley which begins at Cairo. M. Costa accompanies us, and he has written to his father at the Fountains of Moses to say that we shall dine and sleep there on our way back, returning the next day by road to Suez.

" The first part of our voyage went off very well, and at eleven o'clock we landed at the foot of the lofty mountain of Gene'be', situated at the extreme left of a broad bay of which the Attaka formed the extreme right. Opposite to us is the valley wrongly called that of Moses or of the going astray, for the Hebrews never passed that way. It is easy to follow on the map, between Lake Timsah and the basin of the Bitter Lakes, covered at that time by the Red Sea, the march of the Jews when they escaped with Moses from the army of Pharaoh, the very places having, as Linant

Bey has shown, retained the names used in Holy Writ. We are ten leagues from Suez, and as the boats from our steamer cannot go right to shore, we are carried over the intervening distance by the sailors. After reconnoitring the district, we find that the mountain is composed of strata of alabaster, marble, and other calcareous stones, while Linant Bey discovers that behind Mount Gene'be', to the south, there are fields of all the most valuable varieties of marble. We hope, upon re-embarking at two, to reach between four and five the Fountains of Moses, which really are one of the places at which the Hebrews halted after the passage of the Red Sea. But after we had got about half way an accident occurred to the boiler, and though no serious damage ensued we had to spend the whole of the night on board, and instead of enjoying M. Costa's dinner got back to our hotel at Suez at six in the morning."

" December 30, 1854.

" After we had had some breakfast we gave orders for our caravan of camels and dromedaries to be ready for a start, and though my companions were still suffering a little from the fatigue of the previous day and night, we went to our tents, which were put up between the gate of the town and the ruined citadel of ancient Chlysma, upon the sea shore, at the end of the port. It is here that will be the mouth of the Nile water canal, which is an indispensable auxiliary of the

maritime canal. Our encampment, which will remain the same for the rest of our expedition, consists
of three round tents twenty feet in diameter, the first
being for Linant and myself, the second for Mougel
and the young engineer Aïvas, who acts as secretary
to Linant, and who, having been brought up in Egypt,
speaks Arabic like a native, and the third for the
servants. There is a fourth tent which is used as a
kitchen. Some twenty barrels of Nile water are
placed between the first and second tents, and watched
day and night, for upon them our safety depends
during our expedition. Around the kitchen are cages
containing poultry and pigeons, and I am surprised to
find that though these cages are left open during the
day their inmates make no attempt to get away.
There is also a small flock of sheep and goats, thirty-
three camels and dromedaries in the care of fifteen
Bedouins who sleep among them, and a couple of asses
for the use of Mougel, who cannot endure the motion of
the dromedary.

"We agree to start early on the morrow, and M.
Aïvas reads us the memorandum of M. Linant relative
to the levellings executed under his direction in 1853
along the whole length of the isthmus from Suez to
Pelusium."

"*December* 31, 1854.

" Soon after daylight we are ready for a start, and
find the dromedaries ready saddled and crouched down

for us to mount them. To get astride a dromedary is a task requiring no little adroitness, for as soon as you have thrown the right leg over their backs they rise, and those who rise the quickest are the best. The best way is to bring the body slightly forward and make just the opposite motion when the beast stretches out its hind legs. The dromedaries provided for Linant and myself are two very fine animals, and I soon get accustomed to the motion, and other drome- daries are bestridden by M. Aïvas, an Arab effendi, and an assistant-engineer. The Bedouin sheik Jaoudé acts as guide, and has been made responsible for our safety at the price of his life. Our little troop is made up of five messengers, who will carry letters to Cairo, and Linant's ' dusky ' maître d'hotel.

" For three hours we follow the bed of the ancient canal, the two banks of which are still in a perfect state of preservation, and before sunset we pitch our tents in the desert at a spot called Makfar (the Hol- lowed Place), where there is a little vegetation of a meagre kind.

" Opening my Bible, which is always of special interest to read when in Egypt, I am more than ever impressed by it now, for I am drawing near to the region where Jacob and his family established them- selves, and from which, four centuries later, Moses led the Hebrew people out of servitude.

" After reading some passages in the Pentateuch, I went out to admire the beautiful sunset, and was

met by Linant and Mougel, who told me that they
had just seen a luminous meteor shooting from the
east like a rocket, and, after describing a semi-circle,
breaking up into a shower of fire towards the west.
I only just saw the tail end of the phenomenon, so if
it had not been for the evidence of my companions I
should not have ventured to record it. If we lived
in the age of signs and miracles, I might compare
this vision with that which I had seen on the morning
of November 15th, the day when the final decision as
to the Suez Canal was given. This one occurred on
December 31st, the opening day of our voyage of exploring, during which we laid the first foundations of
a union between the east and the west.

" I got M. Linant to read his treatise upon the
ancient geography of the isthmus compared to the
modern, and upon the route taken by the Hebrews,
so as to be able to profit by the impressions which I
had just received from reading some of the principal
passages in the book of Moses."

                                        " *January* 1, 1855.

" We are on the move by five o'clock, and go round
in a westerly direction the basin of the Bitter Lakes,
which formed part of the Red Sea in the time of
Moses, but which is now dried up. One can see
scattered over what was the bed of the sea lumps of
saline composition which look like bits of ice after a
thaw. We go over part of this ground, leaving to

our right swamps in which a traveller would be
engulfed, horse and all, without the least chance of
escape. About half a mile to our left are the Awebet
Mountains, from which good building stone and lime-
stone might be taken.

" We ascend a hill where we see blocks of granite
which have composed what is called the Persepolitan
monument, and one of these blocks is covered with
cuneiform or Assyrian inscriptions, while upon another
is a vulture with outstretched wings, and the ancient
royal wand of Egypt at each corner.

" This monument is supposed to have been erected
by Darius, the Persian conqueror, after his expedition
to Egypt, either to mark a boundary of territory or
to perpetuate the recollection of the reconstruction of
the canal of the Pharaohs, which is attributed to him
by Herodotus. The stone used for it is granite from
Mount Sinai.

" We still follow the desert, leaving the basin of the
Bitter Lakes to our right, and find the ground less
hard than that which we had ridden over the day
before; the sand is not so coarse, and preserves the
footsteps of all the animals which have run over it.
We notice the footsteps of hyænas, gazelles, foxes, and
hares crossing one another in all directions. At four
o'clock we encamp in the valley of the Akram, which
is the name given to a certain kind of bush.

" In the evening our conversation turns upon the
canal from Suez up to the point which we have

reached, and the result of our observations is that the maritime canal, at the entrance to Suez, would receive, in the course of twenty-four hours, a supply of ten million cubic metres of water from the Red Sea tides, and that with this as well as with the waters of the Mediterranean it will be easy to fill the basin of the Bitter Lakes, which is to form an immense reservoir holding two thousand million cubic metres of water. This mass of water will suffice to feed the canal, and will create, when the winds do not force it back, a slight current towards the Mediterranean."

" January 2, 1855.

"We start at six, the wind being cold and unpleasant. We visit a second Persepolitan monument erroneously called Serapeum, and we travel along the dyke of the old canal, which now forms a road. At two o'clock we reach the place where we form our third encampment, near Lake Timsah (the Lake of Crocodiles), against the bir (well) Abdullah. This place is called in Scripture Pi-hahiroth, which in Hebrew means the Valley of Reeds, and it is still called by a similar name in Arabic, *Oued-bet-el-Bouze.* We are in the midst of the land of Gessen (Goshen), and I may here quote the verses from the Bible relating to this country, which we shall visit to-morrow.

" *Genesis*, chap. 46, v. 28.—'And he sent Judah before him unto Joseph, to direct his face unto Goshen; and they came into the land of Goshen.'

"V. 33.—' And it shall come to pass, when Pharaoh shall call upon you, and shall say, What is your occupation? '

" V. 34.—' That ye shall say, Thy servants' trade hath been about cattle from our youth even until now, both we, and also our fathers: that ye may dwell in the land of Goshen.'

" *Gessen* (Goshen) in Hebrew means Pasturage, and it is worthy of note that in Arabic it is *Guess*. In the old Ethiopian language *sos* means shepherd, and this is probably the etymology of Suez. Thus the land of Gessen, or Guess, and the Isthmus of Suez would be denominations applying to the same places. We know now that the dynasty of the Hiksos, which reigned in Egypt signifies the dynasty of Armed Shepherds.

"Chap. 47, v. 4.—' Now therefore, we pray thee, let thy servants dwell in the land of Goshen.'

" V. 6.—' The land of Egypt is before thee; in the best of the land make thy father and brethren to dwell; in the land of Goshen let them dwell.'

" V. 11.—' And Joseph placed his father and brethren, and gave them a possession in the land of Egypt, in the best of the land, in the land of Rameses, as Pharaoh had commanded.'

" *Exodus*, chap. 1, v. 7.—' And the children of Israel were fruitful, and increased abundantly, and multiplied, and waxed exceeding mighty ; and the land was filled with them.'

"V. 8.—'Now there arose up a new king over Egypt, which knew not Joseph.'

"V. 11.—'Therefore they did set over them taskmasters to afflict them with their burdens. And they built for Pharaoh treasure cities, Pithom and Rameses.'

"Chap. 8, v. 22. *The Plague of Flies.*—'And I will sever in that day the land of Goshen, in which my people dwell, that no swarms of flies shall be there.'

"Chap. 9, v. 6. The Sixth Plague. *The Plague of Boils.*—'And all the cattle of Egypt died: but of the cattle of the children of Israel died not one.'

"V. 26. Seventh Plague. *Rain and Hail.*—'Only in the land of Goshen, where the children of Israel were, was there no hail.'

" The inductions which I draw from these passages in the Bible are that the land of Goshen, which will be intersected by our subsidiary canal from the Nile, will become at least as fertile as it was in ancient times, and that its climate is most salubrious, seeing that in our day, as in the time of Moses, the few tribes of Arab shepherds who encamp there are generally exempt from epidemics, despite their coming in contact with the populations of Lower Egypt.

"Chap. 12, v. 37.—'And the children of Israel journeyed from Rameses to Succoth.'

"Succoth in Hebrew means Tents. This place is now called by the Arabs either Oum-Riam (the Mother of

Tents) or Makfar (the Hollowed-out Place where the old canal passed).

" Chap. 13, v. 20.—' And they took their journey from Succoth, and encamped in Etham, in the edge of the wilderness.'

"Etham, which is in fact at the edge of a vast wilderness bounded by the basin of the Bitter Lakes, where the Red Sea then reached, is a spot which we visited this morning, and where the second Perse-politan monument is to be found. The tribe which at certain seasons encamps there is known as the Ethamis.

" V. 22.—' He took not away the pillar of the cloud by day, nor the pillar of fire by night, from before the people.'

" Moses having in his youth killed an Egyptian who was maltreating an Israelite, had to fly to Mount Sinai, where he married the daughter of Jethro, the priest of Midian, whose flocks he tended for forty years. His brother Aaron came to tell him of the death of the Pharaoh during whose reign he had slain the Egyptian, which period corresponds with the length of the reign of Rhamses II., the Sesostris of the Greeks. He was acquainted with the customs of the numerous caravans, which in our own day and at certain places are preceded some distance in advance by bearers of *machallahs* (torches), which produce a pillar of fire at night. They had passed over the fords of the last lagoons of the Red Sea, as is still done by the Bedouins, who avail themselves of the low tide to cross the Red Sea near Suez.

" The generals of Pharaoh, like those of Said Pasha, were unaware of the existence of these fords.

" Chap. 14, v. 2.—' Speak unto the children of Israel, that they turn and encamp before Pi-hahiroth, between Magdol and the sea, over against Baalsephon : before it shall ye encamp in the sea.'

" This place is, of a truth, amid the site of the old lagoons of the Red Sea, the level of which is below that of the sea, and which will become lagoons again when the maritime canal is open. The Israelites, favoured by the tempest which the Bible describes, will have crossed at night, when the tide was low, across the low-lying tract of land between the basin of Lake Timsah and that of the Bitter Lakes, between the long sandbanks which by moonlight would have looked like white walls, and the next morning, the wind having gone down, the Egyptian troops went in pursuit of the Hebrews, and were overwhelmed by the floods or engulfed in the quicksands of the valley.

" Chap. 15, v. 22.—' So Moses brought Israel out of the Red Sea, and they went out into the wilderness of Shur ; and they went three days in the wilderness, and found no water.'

" V. 23.—'And when they came to Marah, they could not drink of the waters of Marah, for they were bitter.'

" In the desert of *Shur* (the desert of Syria, the other side of the Timsah and the Bitter Lakes) is a spring which is marked upon all the maps by the name of *Bir-Mara*, *Mara* signifying *bitter* in the Hebrew as well as in Arabic.

"V. 25.—'But Moses cried unto the Lord; and the Lord shewed him a tree, which when he had cast into the waters, the waters were made sweet.'

"Linant tells us that this practice was taught him by the Arabs of Mount Sinai, with whom he spent a good deal of time, and that in order to diminish the bitterness of brackish water they throw into it the branches of a shrub called arak, a species of thorn.

"V. 27.—'And they came to Elim, where were twelve wells of water, and threescore and ten palm-trees: and they encamped there by the waters.'

"Elim is the place now known as the Fountains of Moses, twelve in number, which we were twice disappointed in our efforts to visit.

"Chap. 16, v. 14.—'And when the dew that lay was gone, behold, upon the face of the wilderness there lay a small round thing, as the hoar frost on the ground.'

"The Israelites exclaimed, 'Man-hu?' which in Hebrew means: 'What is it?' This is the etymology of manna. M. Linant saw the Arabs on Mount Sinai gather up the manna which falls in the morning at sunrise, when the temperature is at a particular point, from the leaves of the tamaris, which grows wild nearly all over the desert we have just traversed as well as upon the other side of Sinai. It is in fact a sort of jelly or white froth, which forms of a morning upon the leaves of the tamaris, and which, melting when the sun begins to shine, falls drop by drop on the ground and forms a paste something like honey. The manna of Sinai is not disagreeable to the taste, and has not the aperient effects of the Sicily manna which is to be found in the chemists' shops.

"The researches of Linant will show that the country in which we now are is in reality the land of Goshen. Having reached our encampment in the Valley of Reeds (Pi-hahiroth) in good time, we avail ourselves of it to compare our observations upon the course of the canal from Suez to Lake Timsah, to study anew the maps and plans of Linant, to discuss the depth and width of the new canal, and to lay down the main lines of the report upon which Linant and Mougel have to agree.

" We are anxious also to form an idea—which can, however, be only an imperfect one—as to the cost. After a great many calculations, we arrive at a total of from six and a half to seven millions. The canal communicating with the Nile from Cairo to the inland port of the maritime canal, and which from that point will branch off into two irrigating and feeding canals, one going to Suez and the other to the Mediterranean port, is estimated at £800,000. All this, as you may imagine, will have to be revised and modified, but not, I hope, much increased, with all manner of documents and plans to bear it out."

" January 3, 1855.

" At eight we start to make an excursion to the ruins of Rameses, and we halt for a quarter of an hour upon the hill where was Succoth, the first station of the Israelites, and we reach the place whence they started on their journey. There can be no longer any doubt as to the site of Rameses since the discovery of the statues which Linant, as you will see, has sketched for me. The hieroglyphic inscriptions which are carved on the blocks of granite, and which have been trans lated, explain that the figures represent Rhamses II. (Sesostris) and his two sons. The ground is strewn with fragments of the ancient bricks the making of which rendered the lives of the Israelites so 'bitter with hard bondage' (Exodus, chap. 1, v. 14.).

" We breakfast in the presence of Sesostris, and return to camp after having followed a part of the ancient canal connecting the Nile and the Red Sea."

" January 4, 1855.

" We start at seven to go round Lake Timsah in a westerly direction. We mark the site of Baalsephon upon a hill at the foot of which the main canal will pass. The sky becomes covered with clouds and the sand begins to whirl around the bushes, and as we are threatened with a tempest similar to that under cover of which the Israelites escaped from the chariots of the Egyptians, we hasten back to camp."

" January 5, 1855.

" It has rained all night, and as the wind is rising still there is no possibility of moving outside our tents, while, looking in the direction from which the wind comes, we see that a violent khamseen is imminent. While we are discussing the question of the canal, a gust of wind more violent than those which preceded it very nearly blew us out of the tent, and upset all our things. The conversation is renewed when the mischief has been remedied, and we calculate that the Sweet Water Canal will be the means of bringing 250,000 acres under cultivation. Just at this moment a messenger arrives with letters from Cairo, one of them being from Admiral Jurien de la Gravière before Sebastopol, while Kœnig Bey sends me a copy of the Viceroy's reply to the letter from Napoleon III. when sending him the Legion of Honour. In this letter the Viceroy speaks very favourably of the prospects of our canal; while another letter from Señor Baguer y Ribas, Consul-General of Spain at Cairo, informs me that Mr. Wilcox, one of the directors of the P. & O. Steam Company, has arrived there, accompanied by a Spaniard, Señor Zulueta, both of whom will, he considers, be inclined to associate themselves in my work."

" January 7, 1855.

" The weather having at last cleared up, we leave Pi-hahiroth for the north of Lake Timsah, which is the central point of the isthmus, and I can at once see how well advised Linant was in proposing that there should be an inland port here. This basin is surrounded by hills, and it forms a magnificent natural port, six times larger than that of Marseilles, and all the more useful because it can easily be placed in communication with the cultivated portions of the land of Goshen and the interior of Egypt, by means of a canal branching off from the Nile. The steamers which cast anchor there will find means of revictualling as well."

" January 8, 1855.

" We begin to descry to our left Lake Mensaleh, formed partly by a rising of the Nile and partly by the waters of the Mediterranean, the shore of which has several breaks between Damietta and Pelusium.

" We halt at noon to breakfast in an oasis, the trees of which have a very pleasing effect upon the eye amidst the boundless desert. I counted twenty-three date-trees in this oasis, which the Arabs name Bir-el-Bourj (the Well of the Tower), there being amid the date-trees a well of brackish water, and upon the hill the remains of what might have been a tower.

" We wait there for the passage of our caravan, and then we wend round the eastern limit of Lake Mensaleh, where we see a quantity of white lines formed by swans, pelicans, and flamingoes.

" We then reach the foot of a hill on which was built the ancient fortress of Magdol, of which the

Bible speaks, and which travellers call Magdolum. While our encampment is being prepared we visit the ruins of the fort. History tells us that this fort was burnt, and we can trace the effects of fire upon the stones, while in the distance to the right we see the shores of Pelusium, where Pompey met his death, and to the left Damietta, where St. Louis landed."

"*January* 9, 1855.

"We make for Pelusium, close to which are the ruins of the modern castle of *Tineh,* which like Pelusium itself signifies *mud,* and at this season the mud itself is covered with water owing to the inundation of the Nile, so we can only contemplate from a distance the ruins of what was one of the most important cities in ancient Egypt. So terminates our first explory, the result of which goes to assure us that our undertaking is practicable, and I hope that the reports of the two engineers who accompany me will prove this to be so.

"We pass the night at the oasis of Bir-el-Bourj."

"*January* 10, 1855.

" The cold is so intense that we walk some distance by the side of our dromedaries, which soon take us on in advance of Mougel when we once mount them, as the keen air has quickened them very much, while Mougel presents a very ludicrous appearance as he comes on behind with his donkey-driver belabouring his mount in the hope of overtaking us, the poor fellow being in mortal fear of the Arabs whenever he is left a little way behind. As soon as Mougel has got up to us we resume our journey by the shores of Lake Mensaleh, instead of crossing the desert as we had done in coming. At five we reach El Guisr, and pitch our tents at the foot of one of the highest downs in the isthmus."

"January 11—14, 1855.

"We return to Cairo, and I find time to draw out a draft of a report for the engineers and read it to Linant and Mougel, who express their approval of its tenour, but suggest that I should strike out the passage in which I propose that, should they differ in opinion, they should give their reasons. But I decline to do so because, with this reservation made, their harmony, upon which I quite reckon, will have all the more weight."

"January 15, 1855.

"At eleven we start at a good pace upon our dromedaries for Cairo. To the left is the chain of mountains which, commencing at the Mokattan and finishing at the Attaka, skirts the road to Suez, while to the right we can distinguish, amid the palm-trees and a line of greenery indicating the course of the Nile, the minarets of Kanka. The morning was a singularly beautiful one, and the view unfolded before me is delightful. We have passed Abuzabel, and we can see the obelisk of Heliopolis, the City of the Sun, where Plato studied for seventeen years the archives of the Egyptian priests. It is a mistake to suppose that this city was the residence of Joseph, the son of Jacob. The dynasty of shepherds, under whose sway Egypt at that time was, reigned at Suez, near Lake Mensaleh, where Potiphar, the first minister of the Pharaohs, also was, as the Scriptures tell us, chief eunuch, a circumstance which pleads in favour of Madame Potiphar and makes Joseph's attitude of reserve all the more meritorious.

" We pass close beside the pretty village of Ma-tarié, surrounded by gardens, in the midst of which is to be seen the Virgin's Tree. We then pass Birket-el-Haggi (the Lake of Pilgrims), where the great Mecca caravan, conveying the sacred carpet to the tomb of the Prophet, meets every year.

" We see before us the massive palace of Abassié, built by Abbas Pasha, and in which there are 2,000 windows, while beyond it, upon the other side of the Nile, are the summits of the two great pyramids which have been in existence not for forty, as is generally said, but for sixty centuries. To the left are the lofty steeples of the mosque built of oriental alabaster inside the citadel by Mehemet Ali, which that great man had arranged should be his burial place. He was well entitled to have for his tomb a spot where he had annihilated the representatives of barbarism and which commanded the whole of the country which he had regenerated. As I ride along I am struck by a coincidence which had not occurred to me before. My father in 1803 was French political agent in Egypt, and Bonaparte had instructed him to pick out from among the leaders of the Turkish troops a man of energy and ability who could be proposed at Constantinople for the hitherto nominal dignity of Pasha of Cairo. Mehemet Ali, who was a native of Macedonia, in command of a thousand Bashi-Bazouks, and who could neither read nor write, had become the friend of my father, who gave him good advice how to deal with the Mamelukes who were hostile to France, and it was Mehemet's name which, at my father's suggestion, was submitted to the Sultan and agreed to.

"And now after the lapse of half a century the son of Count Mathieu de Lesseps, being already the friend of the son of Mehemet Ali, becomes his adviser for the accomplishment of a work which cannot fail to illustrate his reign, and here in this same citadel which witnessed the elevation of Mehemet Ali, following upon the massacre of the Mamelukes, Mohammed Said effects a peaceful *coup d'état* which will complete the regeneration of Europe by announcing to all the Powers of Europe that he has resolved to join the Red Sea to the Mediterranean and open to the whole world the route to India.

"I am convinced that England will profit more than any other nation from the advantages of this

route; but there is no use concealing the fact that
the old egotistical policy of Great Britain receives a
mortal blow, and this is why the partisans of the traditions of old are already in a great state of excitement. I was quite prepared for this, for I had had
better opportunities than any one else, both from
what my father had told me and from what I had
seen myself, of following at various periods their
course of policy in Egypt. Why did they spare no
effort to make General Bonaparte's expedition a
failure? Why, after this, did they protect the
Mamelukes, who were dividing the country, driving
away foreign trade, and condemning the fertile valley
of the Nile to sterility? Why, in 1840, did they induce
the whole of Europe to form a league against France
and Mehemet Ali, the progress made by whom they endeavoured to arrest? Why did they give their support
and advice to Abbas Pasha, that fanatical prince so
opposed to progress, whom Providence removed before he had quite time to complete the disorganisation
and ruin of Egypt? Why simply because there was
a party in England anxious to reduce the Viceroy
to the condition of those Indian rajahs who are
encouraged to lead disorderly lives, until at last
they are so debased and besotted that they have
no alternative except to put themselves under the
protection of England or to sell their states.
Fortunately this opinion is not universal in
England, there being in that land of liberty a

great many noble-hearted and intelligent men who, sooner or later, will carry public opinion with them.

" My letter to Cobden will be useful as a text when the time comes to carry on a crusade against the men who stand by the past.

" After this digression, excited by the reference to the Cairo citadel, let me return to our expedition, which halted for the last time near the tomb of Malek Adel, the cupola of which protects us from the sun. Malek Adel was the brother of the Sultan Salah Eddin (Saladin), who was the reigning Caliph in Egypt during the crusade of Philip Augustus; but Madame Cottin's French romance makes far more of him than the Arab writers, according to whom he did not play any conspicuous part in the history of his day.

" From there we ride on into Cairo, and I again take up my quarters at the palace of the Muçafirs."

Instructions to MM. Linant Bey and Mougel Bey, with reference to the Scheme for a Maritime Canal from the Red Sea to the Mediterranean, and for an Alimentary Canal branching from the Nile.

"CAIRO, *January* 15, 1855.

" The exploring mission with which the Viceroy has entrusted us being just over, I think it well to direct the attention of MM. Linant and Mougel Bey to the principal points which should be present in their minds when they draw up the preliminary project we have agreed to prepare, pending a more complete work accompanied by plans, maps, sections, estimates, and other confirmatory documents :—

" 1. In regard to the entrance from the Red Sea —make it clear whether the present port will be used, what work will have to be done in the way of jetties, &c.

" 2. Indicate the precise direction of the canal from the sea to the ancient basin of the Red Sea, called the Bitter Lakes.

" 3. Explain in what way it is proposed to utilise this basin, and if in passing through it the maritime canal ought to have a continuous bank or none at all.

" 4. Trace the canal to the basin of Lake Timsah, intended to serve as an inland port.

" 5. Works to be executed in order to make Lake Timsah suitable for the object proposed. Give the length of the walls of the quay. In passing through Lake Timsah the canal will have to be much broader than at other points, so that vessels may be able to anchor there, or approach the quay without obstructing the passage. These quays are to be formed as near as possible to the Sweet Water Canal.

" 6. Direction of the maritime canal from Lake Timsah to Lake Mensaleh.

" 7. Works to be executed along the shores of Lake Mensaleh, or in the lake itself.

"8. Will the mouth of the lake upon the Medi-terranean side be the mouth of the ancient Pelusium branch?

"9. Carefully define the character and dimensions of the work to be done in the way of jetties, moles, &c, so as to answer the objections made up to the present time as to the difficulties and supposed impos-sibilities arising from the alluvial nature of the coast and the silting up of the sand at the mouth of a canal to the Mediterranean. This part of the report should be based upon irrefutable evidence and examples.

"10. What is the mass of water which would enter at each tide from the Red Sea into the maritime canal?

"11. What advantage can be derived from the height of the tides, not only upon the course of the maritime canal, but on the basin of the Bitter Lakes and at the mouth of the Bay of Pelusium?

"12. Reckon for the maritime canal upon a breadth of a hundred metres (40 inches each) at the line of low water in the Mediterranean, with power to reduce it to 65 or 70 metres in the few parts where the clearance necessary to make it would be too great. The depth of the water is to be six, seven, or eight metres, calculated, of course, according to the low-water level of the Mediterranean, so that the company may be able to select, with a full knowledge of the cost, whichever depth seems best suited at once to its interests and the necessities of navigation.

"13. Reply to the objections raised as to the difficulties of navigation in the Red Sea and the Gulf of Pelusium.

"14. Make a preliminary *maximum* calculation of all the costs, and indicate the time at which presumably the canal may be open for navigation.

"15. Add to the project of a maritime canal one for a communicating, feeding, and irrigating canal brought from the Nile, starting from a point between the damming of the river and Cairo, and proceeding by way of Ouadee to Lake Timsah, the dimensions of this canal to be so calculated that with its draught of water it will be capable of irrigating about 250,000 acres when the Nile rises. This canal should, when it reaches Lake Timsah, with which it is to be connected, branch off into two parts, one going to Suez the other towards Pelusium.

"16. Examine whether the sands of the isthmus are likely to interfere with the working of the canal, and what use can be made of them by means of the irrigating canal.

"17. Give a maximum estimate of the cost of the subsidiary canal and of the time required to make it.

"18. Point out the character, quality, and site at which are to be had easily, and with little cost of transport, the materials required for the works.

"19. Finally, give an estimate of the minimum

income which might be expected from the maritime
and the Sweet Water Canals.

" I do not for a moment wish to confine their report within the limits indicated by this programme.
Though I have been a witness of the harmony of
their views and of the identity of their conviction as
to the possibility of communicating between the Red
Sea and the Mediterranean by a eanal accessible to
the largest vessels, I beg of them, in the event of
their entertaining opposite views on any subject, to
record the difference of their opinions, and give the
reasons for them.

" Finally, this rough draft should be completed as
quickly as possible and accompanied by an explanatory
map."

The next thing was to go to Constantinople, in
order to obtain the Sultan's assent to the scheme,
and having first appointed my old friend M. Ruyssenaers agent of the company at Alexandria, I found
that the ground was clear.

*Note for the Viceroy of Egypt and for Count Theodore
de Lesseps, Director of Foreign Affairs.*

" CONSTANTINOPLE, *February* 15, 1855.

" Upon reaching Constantinople I found the ground
quite free. If no opinion had been expressed in
favour of the project, nothing had been said against
it. For the two first days I endeavoured to form an
exact opinion of the situation. I had learnt that the

Ministers in general, and Reschid Pasha more espe
cially, *entirely* approved of the scheme, and would be
only too glad to oblige the Viceroy if they could do so
without compromising themselves overmuch.

" But there could be no doubt that they are all here
under the pressure of, not to say dependent upon, the
English Ambassador, whom the public call Sultan
Stratford, or Abd-ul-Canning.

" I assured myself of the support of Baron de
Bruck, the internuncio of Austria, of M. de Souza,
the minister of Spain, of Count de Zuylen, the
chargé d'affaires of Holland, and of several other
personages who were in various ways likely to be of
use to me.

" I heard that Lord Stratford de Redcliffe was personally very much opposed to the scheme, that he had
received no official instructions from his Government,
but that, when the occasion arose, he would act as if
he had, in accordance with his arrogance and deep-
rooted jealousy of all that is French and his incor-
rigible and antiquated British egotism.

" It was doubtful, I was told, whether Reschid
Pasha would dare to shake off the yoke, intolerable as
it was beginning to be felt.

" M. Benedetti, our chargé d'affaires, whose local
experience, tact, and prudence were to stand me in
such good stead, saw that it would not do for him to
put himself too forward, but he promised me to do all
he could to facilitate the success of a negotiation by

which, as he knew, the Imperial Government set great
store. From the very first he instructed M. Schefer
to pave the way for an interview between Reschid
Pasha and myself, and this interview took place at his
house on the Bosphorus on the morning of the 12th.
I handed the Grand Vizier a letter from the Viceroy,
and a Turkish translation of my memorandum and of
the firman of concession for which the Sultan's assent
was sought.

"After carefully reading these documents, the
Grand Vizier and myself conferred for two hours, and
I felt that at the expiration of that time I had made
some impression upon him by using the arguments
which I have already quoted. I added that I was
only there as a friend of the Porte and an agent of the
Viceroy, not as an agent of the French Government,
which had entrusted me with no mission of any kind.

"The next day, during a grand banquet given by
Aali Pasha, the Minister of Foreign Affairs, previous
to the departure of Baron de Bruck, Reschid Pasha
ventured to speak to Lord Stratford de Redcliffe about
the object of my visit to Constantinople, and encountered the resistance which we had anticipated.

"It became necessary, therefore, to press for decision and act vigorously, so as to forestall the demands
of the English Ambassador upon the Sultan. The
Austrian Internuncio, to whom I reported how matters
stood, assured me that I might count upon his private
and official support, and that though his successor, who

was expected in a few days, would doubtless bring special instructions, he should not hesitate in the meanwhile to act on his own judgment and counterbalance the influence of Lord Stratford de Redcliffe."

While the French Embassy was asking the Grand Vizier to obtain me an audience of the Sultan, I drew up the following note for the members of the Imperial Council.

" CONSTANTINOPLE, *February* 19, 1855.

" It would be superfluous to dwell in detail upon the immense advantages of the piercing of the Isthmus of Suez. The only obstacles which seem to stand in the way are the personal objections raised by a foreign representative, objections which, if they were allowed to prevail, would inflict a moral blow upon the highest authority in the land. I feel confident that this obstacle will not be allowed to prevail against the wishes which I have been charged to express, with all respect, upon behalf of an enlightened Prince who, as is his duty, does an act of deference to his sovereign, whose faithful and devoted vassal it is his pleasure to prove himself."

The Sultan granted me an audience and received me very graciously, and I was about to embark upon my return to Egypt, when I learnt that the influence of Lord Stratford de Redcliffe had been strong enough to deter the Sultan from sanctioning the project out-

right, and so I determined to stay a few days longer
and see what could be done.

The day following, at a dinner at the British Embassy to which I was invited, I sent Lord Stratford
de Redcliffe the following letter:—

" CONSTANTINOPLE, *February* 26, 1855.

" I hasten to communicate to you the documents
which, in accordance with the wish expressed by you,
will enable you to form an opinion as to the enterprise
which has brought me to Constantinople.   I venture
to hope that I shall no longer have to fight against the
powerful opposition of the honourable representative
of Great Britain.

" Your Excellency was pleased to tell me that you
were anxious for information on the subject, and that
up to the present time you had only given a personal
opinion.

" The question has been submitted in due course to
the Sublime Porte without any sort of foreign intervention.   It would not be within my province, as the
agent of Mohammed Said, to place it upon another
ground, as your Excellency suggested.   The Viceroy
of Egypt was at liberty to place it upon this ground
and keep it there.   Just as he was unwilling to give
it a purely French or Austrian complexion, in the
same way he would not assent to give it an exclusively English aspect by transferring the discussion
of it to London, and letting the solution of it depend

upon one Government.   He is anxious that this affair of the Suez Canal should retain, above all things, its Egyptian and Ottoman initiative.

" Your Excellency is too enlightened a patriot and attaches too much importance to the alliance between our two countries—an alliance of which I am proud to be one of the warmest partisans—to allow a question of antagonism, in which it would be deplorable that the *amour-propre* of our two Governments should be involved, to arise in this connection.

" Your Excellency will not allow it to be said that England, which with justice declares that she has only drawn the sword against Russia in the interests of civilisation, of the freedom of the seas, and of the independence of Turkey, should be the only Power to place difficulties in the way of a work which essentially favours the realisation of principles which should be the consequence of the Anglo-Austro-French alliance, and which will assure the pacification of the East.

" I am pleased, my lord, to have had this conversation with you.   It has had the effect of destroying impressions which, I do not hesitate to say, I had erroneously formed.   I ask your permission to renew the conversation, and with that view I will call at the English Embassy about one to-morrow.

" P.S. The Viceroy has just informed me by a letter from Alexandria, under date of February 17, that up to that time Mr. Bruce had made no communication to him on behalf of his Government."

Reply of Lord Stratford to M. de Lesseps.

(Private.)

" Hôtel d'Angleterre, *February* 27, 1855.

" I write you at an early hour, not only to acknowledge the receipt of the documents which accompany your note, but also to ask you to defer till another day your proposed visit. Engagements which I cannot put off make it impossible for me to avail myself of your obliging proposal to-day.

" You are right in supposing that I am anxious for information, and especially in respect to this or to any other great enterprise which closely touches the interests of more than one State, and which, while being theoretically so seductive, causes a great division of opinion from the practical point of view.

" You are too enlightened and experienced to complain if I do not say more. The various considerations which you have touched upon in a manner at once delicate and flattering to myself, are at the same time of too high a political order to be entered upon here.

" In a position such as mine, personal independence has its limits, and cannot but yield at times to official eventualities."

To Comte Th. de Lesseps, Paris.

" Smyrna, *March* 3, 1855.

" My note to Lord Stratford de Redcliffe was translated into Turkish, and submitted on the 27th to the

Council of the Sultan, to whom my letter of the 24th, addressed to Reschid Pasha, had already been communicated. These communications produced the effect which I had hoped. It was proved, upon the one hand, that there was no reason for expecting any fresh explanations from the Viceroy, and, upon the other, that Lord Stratford de Redcliffe had received no instructions from his Government, which has nevertheless been in possession of Mohammed Said's determination for the last three months. It was also proved, by my latest correspondence with Egypt, that Mr. Bruce had not, up to the 17th, made any official representation, despite the intimation to the contrary of the English Ambassador.

"I was informed that Lord Stratford de Redcliffe would strongly urge the Grand Vizier to await before replying conclusively to the Viceroy the instructions which he had applied for to London. The last phrase in his note of the 27th, referring to ' personal independence having to give way to official eventualities,' gave me to understand that his opposition would not be so easily disposed of. He knows that his position is threatened in England, and that several members of his Government are far from favourable to him. If he does not go so far as to believe that the absence of instructions is a trap set for him, he is not at all events inclined to allow an act to be accomplished which he thinks would be unfavourably received in England. Moreover, in view of the disasters which the English arms have sustained in the Crimea, he is stiffening himself against anything which would tend to show that his omnipotent influence is waning, and this feeling, which one cannot at bottom bring oneself to censure, adds just at this moment a fresh force to his British exclusiveness.

" I was not therefore surprised to learn that on the morning of the 28th, upon which day the Council was to decide finally as to the Isthmus of Suez question, he had had a conference lasting three hours with the Grand Vizier, and that, in view of the impossibility of obtaining a negative solution from the Council, they had agreed upon a plan to gain time.

" Reschid Pasha is supposed, rightly or wrongly, to be the noble lord's humble servant, and to be afraid that he will lose his place when the influence of which in private he complains very much, and says that he should like to shake himself free, declines. But, as a matter of fact, he always submits to the domination of which he complains, and which is becoming intolerable for the credit of France in the East. It would be unwise to put too much faith in his protests, or to be led astray by the approaching despatch of his son as ambassador to Paris. He seems to me to be of too un-decided a character to raise up Turkey out of the abasement into which she has fallen, and to make a good use of the elements of vitality which, in my opinion, she still possesses. The result of the understanding between him and the English Ambassador was that the

Council deferred settling the question upon the pretext of the nomination of a commission of three, who,
without concerning themselves with the Suez Canal
itself, were to examine with me in detail the clauses
of the Egyptian firman. This firman, drawn up under
the inspiration of the Viceroy, almost with his own
hand, one may say, already approved by the Cairo
Divan, and communicated to all the European Cabinets, was the indispensable basis of the undertaking,
and was not even under consideration. I could not
follow my opponents upon this ground, and I had,
moreover, been specially instructed not to discuss it,
the Viceroy being quite resolved not to allow an act of
deference towards his suzerain to be converted into a
precedent which would accustom the Sublime Porte,
often subject to the pressure of fanaticism or foreign
influence, to hamper the action of the Egyptian internal administration. That intelligent Prince is
anxious to set his country a wholesome example, and
to create for himself resources as essential to the prosperity of the empire as they are to the real interests
of the empire. Reschid Pasha seems to me to understand the danger of the course he was about to enter
upon, and as he is very anxious to remain in the good
graces of the Viceroy, he wants to cast all the responsibility upon the Council, behind which he is accustomed to shelter himself. I did not hesitate to tell
him that this would not suffice, and that he would, in
my opinion, be held entirely responsible not only by

the Viceroy, but by the Governments of France and
Austria, both of which, as he knew, were in thorough
sympathy with the enterprise. I added that I had
made up my mind to return to Egypt, where the
Viceroy was awaiting me to prepare the organisation
of his scheme, pending the sovereign's authorisation
of it, about which there could be no doubt in the end,
and I congratulated myself on having, during my stay
in Constantinople, contributed to define the question,
while I carried with me the conviction that the Sultan
and his advisers were well disposed towards a work
the realisation of which seemed to me certain. I promised to convey to the Viceroy the intimation of these
friendly dispositions, which would, I hoped, be soon
followed by more practical results.

" In fine, my impressions, and those of the French
Embassy, as to the date, more or less immediate, at
which we shall obtain our aim are not unfavourable.
The expediency of my departure was concurred in by
Benedetti, who, I am glad to say, did all that it was
possible for one in his position to do."

To the same.

"Smyrna, *March* 10, 1855.

"My departure from Constantinople of course led
to the dissolution of the Commission, which had not
had time to meet. As the Turkish Government
naturally does not like to admit that the delay
deemed necessary before the Sultan comes to a deci-

sion is caused by the fact of Lord Stratford de Red cliffe having applied for instructions from London, I have just been informed that upon the proposal of Reschid Pasha the Divan has reverted to its original resolve of asking for some further explanations from the Viceroy. The Austrian packet on board which I am about to embark is doubtless conveying a second letter from the Grand Vizier to the Viceroy. My letters from Stamboul tell me that the British Ambassador has been endeavouring to alarm the Porte by dwelling upon the importance which this canal would give to Egypt and to the Viceroy, to the ideas of independence which it would foster, and to the danger of ships of war passing through. All this is very dastardly, and seems to show, among other things, that England, more than any other Power, is anxious to keep Turkey weak, in order to maintain the idea that she alone is capable of protecting her."

To the same.

"Saidie (on the Nile), *March* 18, 1855.

"A second letter dictated by Reschid Pasha to Kiamil Pasha, which the Viceroy only got yesterday, repeats in an aggravated form the original objections. This step, which is in contradiction with what the Grand Vizier told me, and with the message of which I was the bearer, is evidently the result of a plan arranged by him and Lord Stratford. It has made

Said Pasha very angry, and has only served to strengthen his resolve.

" I am very satisfied with the report of Linant and Mougel, though it is not quite finished. It is so clear and logical that it must carry conviction to every mind."

To M. Hippolyte Lafosse, Paris.

" CAIRO, *March* 22, 1855.

" I asked you, through Madame Delamalle, to express my sincere thanks to M. Thiers for the way in which he has received the news of the Viceroy having granted the concession for the canal between the two seas. M. Thiers has always possessed in the highest degree the sentiment of patriotism, and one can always tell, when one hears him talk, or when one reads his writings, that his heart has not been desiccated in the wear and tear of business. It was he who twenty years ago entrusted to me the conduct of the French Consulate in Egypt. It was during that period that I made the acquaintance of the young prince, now Viceroy, and formed with him that intimate friendship which has procured for me the honour of being selected to carry out this great work. It would be a work beyond my strength if I could not count upon the sympathy and assistance of the men whose opinions carry universal weight with them. I have been anxious to say to M. Thiers how delighted I should be if he would advocate the cause of the canal in his correspondence and conversation with English statesmen, over whom he exercises so deserved an influence. Give him, if you please, the enclosed copies of my letters to the representatives of Great Britain in Egypt and Constantinople. They will prove to him that our only difficulties proceed from England. It is upon England that we must concentrate our efforts, and if M. Thiers, with his wonderful talent for explaining his views, and with the grandeur which they always possess, will take up my cause, his opinion will carry great weight with it.

"There is one point of view which will not have escaped his notice. The opening of the Suez isthmus will be a very powerful safety valve for the boiler of European revolutions. In 1848 we had evidence of how necessary it was to find occupation for an exuberant population, and to provide some useful employment for the turbulent energies of those whom a rapid increase in the number of inhabitants had left in enforced idleness. It is, therefore, in the interests of all the nations of Europe to favour the junction of the two seas, which offers so vast a field to their present tendencies towards speculation and locomotion.

"England does not like to confess the motives for her opposition; but she must make up her mind to the fact that she can no longer claim the monopoly of the trade of the world, nor supremacy in all waters.

" Our continental wars and the weakening of all
other navies have enabled her to establish well-
selected points of vantage in every sea and to hamper
the trade of other nations. But this is no longer
possible. The dream of universal dominion has passed
away, as the empire of the first Napoleon showed.
If a nation, however powerful she might be, attempted
to debar a means of communication which, by right
of the respect due to the capital subscribed, would be
the common property of all nations, she would very
justly be banned by public opinion and would ultimately have to withdraw her pretensions. It will be
wiser and more profitable for her to abandon her
exclusive ideas of omnipotence and enter into partner-
ship with other nations. Her share will always be
the largest when it is a question of trade, industry,
and navigation.

    *   *   *   *   *

" You ask me upon what bases I propose to place
the financial part of the company. I have, upon that
point, only one principle, very firmly fixed it is true,
but the means for carrying out which must be left to
time and consideration. My object is that, *in all
countries the largest possible number of small shareholders shall enjoy the fullest possible advantages.*

" Suppose I was to come to an arrangement with
ten large bankers to make the concession over to
them, what would happen ? They would propose to
divide so many millions; they would then distribute

to the 'vile multitude,' without spending a penny and with a high premium, five hundred franc shares, taking care to let it be known that, as is very likely to be the case, these shares will one day yield an interest of 20 or 30 per cent.

" Why should we not go direct to the public? The two latest French loans show what can be done with small capitals. When a principle is sound, its consequences are incalculable. You will say that the Canal Company will not inspire the confidence which a strongly-constituted Government enjoys. My answer is that the junction of the Mediterranean and the Indian Ocean, teeming with such immense results, can be put so clearly before the public that the most prejudiced will be convinced that no speculation could offer a better chance of profit to those who take part in it, and that when it is found that the cream of the profit has not fallen a prey to a few brokers, there will be no lack of shareholders. When they see that the enterprise is entered upon without any preliminary corruption or promises, when they know that not a stiver has been spent in trying to obtain the concession, either in Egypt or Constantinople, and that, thanks to the generosity of the Viceroy, the preliminary investigations have cost nothing, they will regard all this as the best guarantee for the future.

" You reproach me for not having had recourse to your cash-box, which is open to me with a disinterestedness very consolatory to those who believe, as I do, that in this world there is, as a rule, more good than evil. You are aware that I have got together a certain number of friends, each of whom has paid in a sum of £200 to a common fund intended to defray the preliminary expenses. It was the privilege of myself and my family to lead the way. But for the royal hospitality of 'my prince,' you would be right in supposing that my expenses would be pretty heavy.

"I am obliged to you for having delivered my message to Barthélemy St. Hilaire ; I hope that his beloved Aristotle will not carry the day, especially as he would not need to give him up if he lent me his aid." *

To H.R.H. the Duc de Brabant (now King of the Belgians).

"CAIRO, *March* 23, 1855.

"I have the honour to submit to your Royal Highness the documents relating to the Suez Canal, which you expressed a desire to see, and I venture to hope that you will call the attention of His Majesty the King of the Belgians to the importance of the scheme, and to the interest which the enlightened prince who governs Egypt attaches to it. At a time when we

* Note of the Translator.—M. Barthélemy St. Hilaire, whom M. de Lesseps induced to come out to Egypt with him, was busy upon his famous translation of Aristotle, which is a work of great erudition and value.

have reason to fear that the undertaking will not be appreciated as it deserves to be by some few statesmen in Great Britain, the opinion of the king your father, whose great experience and wise spirit of conciliation have often rendered such service to Europe, may be of immense weight.

"His Highness the Viceroy, in a reply to a letter from Prince Metternich the other day, requested him to thank the Emperor of Austria for the official support which his internuncio at Constantinople had given to this enterprise, as one of universal benefit, and as one which could only be opposed out of narrow and egotistical motives. I beg of your Royal Highness to accept my thanks and my gratitude for the encouragement you have afforded me."

Confidential Memorandum to His Highness the Viceroy.

"BOURAJAT (LOWER EGYPT), *March* 26, 1855.

"Following upon a conversation between the Emperor Napoleon and the English Ambassador, in the cause of which the Emperor warmly advocated your Highness's scheme for piercing the isthmus of Suez, the English Ambassador had an interview with the Minister of Foreign Affairs, and gave it to be understood that instructions had been sent to Constantinople to temporize only, and not to compromise the political situation of England with regard to France.

"It would appear, therefore, that the English Government, as we imagined, is not at all inclined to compromise itself if the Sultan gives his sanction to the scheme.

"It seems to me, in consequence, expedient that the firman containing my powers for constituting a Universal Company, which has up till now, at my desire, remained in your Highness's possession, should be handed over to me.

"Your Highness is aware that the firman relating to the railway which England had asked for, under Abbas Pasha, had been granted in Egypt, without any previous authorization being demanded from Constantinople. England then maintained that such an authorization was not necessary. Upon the present occasion, you have deemed it well to act more guardedly, and you have done right; but prudence has its limits. Your Highness is convinced, moreover, that I shall not compromise you in any way, and that I shall make only a discreet use of the powers granted me. You have been pleased to recognise the fact that my mission to Constantinople had placed the question in a proper light, and had produced a good effect. We must now lose no further time. France, Austria, and the other Powers, support your Highness's scheme. England cannot oppose it outright; all she can do is to endeavour to gain time. It would be dangerous to let her do this. I repeat, then, that the situation is excellent to enable us to do what we were agreed upon, viz., to continue our onward progress, while your Highness keeps insisting upon the Porte giving its decision. It will be desirable that, in addition to the powers conferred under your Highness's seal, which I will retain in my possession, and only make use of in Europe when I have received permission to do so, I should be the bearer of replies to the special objections offered by Reschid Pasha and Kiamil Pasha."

To Count Th. de Lesseps, Paris.

"ALEXANDRIA, *April* 5, 1855.

"My last letter of March 28th announced my departure from Alexandria to go and meet the Viceroy on the Nile. I met him at Kaferleis, and we spent the evening on his yacht, returning by rail the following day to Alexandria. As soon as I saw him he began to complain very strongly of the attitude assumed by his brother-in-law, Kiamil Pasha, and the Grand Vizier, whose letters he had just read. He says that the most fastidious and exaggerated arguments have been used to alarm him and to intimidate his courage, which, as I perceived, was beginning to waver. He had even been threatened with the wrath of England, whose fleet, once the Black Sea question was settled, might come and attack him.

"He was told that he was very foolish to throw himself into the arms of France, whose Government and whose agents were very unstable, whereas the

English agents, on the contrary, are always backed up and supported, their rancour being, therefore, very dangerous. He was further told that the internal tranquility of France and her external influence are at the mercy of a pistol shot to which the Emperor Napoleon may at any moment succumb; and, lastly, he is warned that if he persists in his scheme, he will lose the good graces of the Sultan. I repeat what the Viceroy said word for word. Anyone who knew his character would be aware that such a system of underhand intimidation would produce the very contrary effect to that desired.

" Reschid Pasha, with the aberration produced by the terror which Lord Stratford de Redcliffe causes him, had not reckoned that Said Pasha had sufficient confidence in me to disclose to me all his unworthy devices.

" It must be allowed that the conduct of England in this matter has been pitiful in the extreme. One can understand that Lord Stratford de Redcliffe, in his conversation with me should have sheltered himself behind official eventualities, though such an attitude, suitable enough at the time of the Egyptian firman being first granted, was scarcely applicable four months afterwards. But to say now that he is awaiting instructions is, as my cousin E. de Lesseps wittily put it, equivalent to asking for soup after dinner is over.

" Is it possible, then, that the English Cabinet, not seeing its way to oppose the project upon grounds that can be openly avowed, is reduced to offering, through the mouth of its ambassador, who is supposed to be airing merely his personal opinion, an opposition as brutal as it is tortuous? Such an opposition, un-worthy of men who call themselves statesmen, could only serve to delay a work which cannot be put back, to discredit a Government and eventually to immolate an able deplomatist who is destined to be sacrificed to the good harmony of two nations whose alliance is not always profitable to our good faith.

" I naturally made these observations to the Vice-roy, and he asked me if I had heard of the arrival at Alexandria of a Turkish general from Constanti-nople, and if anything had transpired with regard to his mission. I told him that rumour had it that he had come to ask for money, horses, and grain.

" When we arrived at Alexandria, the Viceroy sent for this envoy, who was a general named Reschid Pasha. He handed a letter from Riza Pasha, the Sultan's Minister of War, appealing to the Viceroy's generous feelings, and asking him, in view of the ex-treme gravity of affairs, for an extraordinary subsidy in the shape of horses, mules, and grain. Two days after this, the Prince called me on one side and, with a very satisfied air, spoke as follows :—

" 'I replied to Riza Pasha that if his demand had come to me through the Porte, with which I had every reason to be dissatisfied just then, I should have refused point blank ; but that desirous of making myself agreeable to the minister who had shown himself more favourable than any of his colleagues to the making of the Suez Canal, and who had not scrupled to advocate it in council against an opposition of which *some others* stood so much in dread, I was very much disposed to send him what he asked for ; but that as we must all of us look after our own interests, I should, for the present, only get the horses, mules, and corn together, and hold them ready to send back by the vessel which brought the Sultan's sanction to my scheme for making the canal.'

" The Viceroy also wrote a second letter to his brother-in-law, Kiamil Pasha, rebuking him in very round terms for his threats as to what the English fleet would do, and he displayed in the whole of this correspondence a ready wit and firm will, upon which I offered him my sincere compliments. He himself dictated the letter to Riza Pasha, in the presence of the envoy, and this will show that when I spoke in the name of the Prince at Constantinople I knew what his real wishes and views were.

" The letters which I received from all quarters prove how sympathetically the project of the Suez Canal is received in Europe. This is how M. Guizot writes to Count d'Escayrac :—

" ' I am very desirous that the canal should be made, chiefly for the benefit of the civilised world, and in a minor degree out of *amour-propre*. It will be the realisation of one of the designs which I have, I will not say dreamt of, but in a measure foreseen and begun.   The present Viceroy, by carrying out this great enterprise, will confer high honour upon himself, and will elevate in no small degree the *rôle* of Egypt in the affairs of the world.   I cannot say, nor can any one else, what will be the fate of the Mahometan East as a consequence of the efforts which are now being made—whether to maintain it, or to transform—but, in any event, the great canal from the Mediterranean to the Suez will transform the relations of Europe and Asia.   This is a result which is worth working for, and which may be attained amidst the storms and obscurities of the war now in progress.' "

To the same.

" ALEXANDRIA, *April* 8, 1855.

" M. Baude, in the *Revue des Deux Mondes*, advocates the roundabout route for the canal proposed by Talabot, and he bases his argument for it upon the statement that the delta and alluvial deposits on the east coast of Egypt are due entirely to the mud from the Nile, which will block up the entrance to the canal at the Mediterranean end.   This is an erroneous view, which the subjoined report of Linant Bey and Mougel Bey will serve to dispel ; and it will, I may add, be much easier and less costly to execute the works required for preventing any silting up of the kind than to make the canal by the roundabout route, which is two hundred miles long, and which would necessitate from eighteen to twenty-four locks and a very

light draught of water. The subjoined report I am
having translated and reprinted in several languages.

Preliminary Report of MM. Linant Bey and Mougel
Bey, dated Cairo, March 20, 1855.

" The Isthmus of Suez is a narrow neck of land,
the two extreme points of which are Pelusium and
Suez. It forms within a space of from thirty to forty
leagues, a longitudinal depression, resulting from the
intersection of the two plains, descending by a gentle
slope, the one from Egypt, the other from the first
hills of Asia. Nature herself seems to have traced
this line of communication between the two seas.

" The geological conformation of the soil leads one
to believe that in early days the sea covered the valley
of the isthmus ; for there are still several vast basins
there, the largest of which, called the Bitter Lakes,
preserves evident traces of the waters of the sea.

" This basin and that of Lake Timsah will, there
can be no doubt, be of immense value in the formation
of a canal. The Bitter Lakes, to begin with, present
a natural passage which will require no cutting, and a
reservoir of water for feeding it with, and superficies
of 330 million square metres. Then, again, Lake Timsah, situated half way between Suez and Pelusium,
becomes, in the event of the direct route being taken,
the inland port of the canal, at which ships will find
all they require in the way of revictualling or repairs,
and which might, if needful, be a depôt for their

merchandise. It seems certain that from Suez to Pelusium, the excavation will be made in loose earth,
which can be moved by hand down to the water-line,
and with dredges down to the bed of the canal.

"Some people are afraid that a canal cut through
the isthmus would soon silt up, and would, therefore,
be so costly to maintain, that it would have to be
abandoned after it had been made. This objection is
refuted by what we saw in December and January;
for we could trace the encampments of the engineers
who were at work in 1847, and, to go back many
centuries, we may add that the banks of the ancient
canal of the Pharaohs and the Caliphs are still visible.

"No doubt the tropical rains of twelve centuries
have formed ravines through these banks, and in
places have filled them, but nowhere are they buried
beneath the sand, and there are still to be seen upon
the surface vestiges of antiquity several thousand
years old. It is only upon one part of the line of
the canal, as we approach Lake Timsah, that the
sand banks appear to undergo changes of shape
rather than of position. All the sand hills which
form a chain between the extremity of the lake and
Pelusium have long since been settled permanently in
their places by the different plants which have grown
up beneath the influence of the moisture and the heat.

"The question which remained to be solved was
that as to the mouth of the canal, both at the Mediterranean and the Red Sea end.

" We had to consider whether the running out to
sea of a double jetty thirty feet deep, with a canal
between broad enough and deep enough to admit the
passage of the largest vessels, would present insurmountable difficulties. We came to the conclusion
that there was nothing to prevent the establishment
of these jetties, adducing in proof the Cherbourg Jetty,
which is more than $2\frac{1}{3}$ miles long, and goes down
nearly 50 feet into the water; the Plymouth Mole,
which is nearly seven-eighths of a mile long, and is 36
feet deep; and that of Lion Bay, Cape of Good Hope,
which is 5 miles long and more than 50 feet deep.

" All of these works have been attended with difficulties, arising from the force of the current and the
depth of the water, which would not occur here. It
has been asserted that the coast at Pelusium was subject to the alluvial deposits of the Nile, and that in
these parts the sea was charged with such thick mud
that it would soon block up the entrance to the canal.
But we know as a matter of fact that Pelusium, or
rather its ruins, is the same distance from the coast
that it was in Strabo's time, 50 years B.C., that is to
say, rather less than two English miles.

" On the Suez side the process of silting up has
got to be a very slow one, for when the plan of the
harbour was taken in 1847 the soundings corresponded
almost exactly to those of the French expedition in
1799, and both of these tally with Commodore
Moresby's chart of the Red Sea.

" So that all that need be done on this side of the isthmus would be to make two jetties, to form a canal, and take it up the gulf to a point where there would be sufficient water for navigation.

" The roadstead of Suez is protected from all winds, except from the south-east, and the ill-effects of this might be guarded against by prolonging the eastern jetty to the south. Moreover, even as it is, all the vessels which come into the Suez roads are quite safe in bad weather, and the corvette store-ship of the East India Company, which has been there for two years and a-half, has never sustained any damage.

" Having thus ascertained the possibility of making a canal through the isthmus, it is essential to show Egypt can be put in communication with the maritime canal. Near Lake Timsah the longitudinal depression of the isthmus is joined, at right angles, by another and not less remarkable valley, which is called in Arabic Ouadée-Tomilat. It is at present an uncultivated desert, but this desert was formerly the fertile land of Goshen, and the valley receives, throughout its whole extent, the overflow of the Nile's lateral canals, and seems thus to furnish a natural line of communication between that stream and the maritime canal.

" Our proposal is to cut through this valley a canal which would serve not only to irrigate the soil, but for internal navigation, while it would also be useful for conveying sweet water to the men employed on the isthmus.

"There would be two secondary branches of the Sweet Water Canal above Lake Timsah, one towards Suez, the other towards Pelusium. The expenses of the project are estimated at seven and a-half millions.

"The last question examined by us is as to whether the capital invested in this work would yield a fair return, and this, leaving out of consideration all general considerations as to the rapid extension of traffic in all parts of the world, can be answered in the affirmative upon more special grounds according to the most recent statistics. The total value of the exchanges between Europe and North America upon the one hand, and the countries beyond the Cape of Good Hope and Cape Horn on the other, exceed a hundred millions sterling, and this total is certain to be still higher by the time that the canal is open. These goods represent, at the very moderate estimate of £24 a-ton, six million tons, and it is fair to imagine that nearly the whole of this freight will, in a very short time, go through the canal. But if we take only half of it and put the charge at ten francs a ton, we shall find that there will be still a great saving for ships using the canal; and we are so convinced that the above estimates will be rapidly exceeded that we suggest that the company shall insert in its statutes a clause providing that the rates of charge should be lowered as soon as the dividend reached 20 per cent., so that the trade of the world should have its share in the advantages of this great and useful enterprise."

To Madame Delamalle, Paris.

"ALEXANDRIA, *April* 21, 1855.

"You tell me that several financiers are trying to put themselves forward in connection with the canal; but what makes me so independent of them is that the position is that of being charged with the exclusive powers of the Viceroy, whom I have always at my back, as being after all master in his own household, whereas if he had given me the concession before the formation of the company, he would, so to speak, have abandoned his rights, and I should not have been so strong to resist the importunings of speculators and governments.

"When I conceived the idea of asking for powers instead of a concession, I did not know that the same thought had occurred to Prince Louis Napoleon at the time when he was bestowing a good deal of attention upon the means of making the Interoceanic American Canal. In 1842, while a prisoner at Ham, he gave a great deal of consideration to this question, and he afterwards asked an officer of the French navy who was starting for Central America to examine the ground and let him know whether the scheme was a practicable one. The officer did so, and his report was embodied by the Prince in a very interesting pamphlet, which was published in London, though only a very few copies were printed.

" In 1846, the Prince received while still at Ham a letter from M. de Montenegro, Minister of Foreign Affairs in Nicaragua, who officially conferred upon him, ' all the powers necessary for organizing a company in Europe,' and further informed him that the Government of that State had determined that the new route should be called the *Canale Napoleone.*

"The French Government did not even reply to a letter from the Prince, who asked that he might be allowed to leave for America to undertake this mission, and I have reason to believe that he was on the point of going there from London when the Revolution of 1848 opened the gates of France to him.

"Those who, like myself, did not rejoice at his accession or vote for the Empire, cannot fail to perceive, when travelling in foreign countries, how much he has raised the name of France, and must admit that the good sense and instinct of the people were better to be trusted than their own feelings of repugnance.

"I have in my hands a memorandum which the Prince wrote in English, and of which I translate you a few extracts :—

" ' Central America (if I substitute Turkey the Prince's argument will hold good of the Suez Canal) can only hope to emerge from its languor by following the example of the United States, that is to say by borrowing hands and capital from Europe.

" ' The prosperity of Central America concerns the interests of civilisation at large, and the best way of labouring for the good of humanity is to break down the barriers which divide nations, races, and individuals.   It is the course indicated to us by Christianity, and by the efforts of the great men who have appeared at intervals upon the world's stage.   The Christian religion teaches us that we are all brothers, and that in the eyes of God the slave is the equal of his master; so in the same way the Asiatic, the African, and the Indian are the equal of the European. Upon the other hand, the great men of the earth have, by means of the wars which they have waged, mingled together different races, and left behind them some of those imperishable monuments, such as the levelling of mountains, the clearing of forests and the canalizing of rivers, which, by facilitating communications, tend to bring nearer together and to knit in friendship individuals and nations.   War and commerce have civilised the world.   Commerce is still following up its conquests.   Let us open a new route for it.   Let us bring Europe closer to the peoples of Oceania and Australia, and enable the latter to share in the blessings of Christianity and civilisation.

" ' In order to carry out this great enterprise, we appeal to all men of religion and of intelligence, for it is one worthy of their zeal and sympathy.   We invoke the assistance of all statesmen, because every nation is interested in the establishment of new and easy communications between the two hemispheres; and, finally, we appeal to capitalists, because, while sharing in a glorious enterprise, they have the certainty of deriving great pecuniary advantages.'

"I am about to decide in the course of a few days with the Viceroy about my return to France."

Confidential Note to the Viceroy.

"ALEXANDRIA, *April* 28, 1855.

"As a guide to our conduct with regard to the canal, I send your Highness confidentially the private information which has reached me from Paris and Constantinople.

"Let me begin with Paris. The first intimation of your Highness's project was conveyed to M. Drouyn de Lhuys, the Minister of Foreign Affairs, in a telegram from Marseilles on December 13th. M. Thouvenel, the Political Director of the Ministry, sent word of this to my brother, the Director of Commercial Affairs in the Ministry. The next day M. Drouyn de Lhuys received a visit from Lord Cowley, the English Ambassador, who came in a great state of mind to ask for explanations as to what was being done in Egypt, and to inquire whether there was any previous understanding between the French Government and myself. M. Drouyn de Lhuys told him the simple truth, when he said that he was entirely ignorant of what was being done in Egypt, that he did not see

me before I started, and that it was a well known fact
that since my mission to Rome, I had had no dealings
with the Emperor or his Government. He added,
with much dignity, however, that if the report which
he had just heard for the first time turned out to be
a fact, he should personally be very delighted, and
should be fully prepared to support the undertaking.

" Lord Cowley then addressed himself to the Emperor, whose attitude of reserve was interpreted as
being favourable to the views of the English cabinet.
One of my friends, feeling somewhat uneasy on that
score, instituted inquiries, and he writes me : 'The
Empress asks me to say that, upon her again ques-
tioning the Emperor, the latter told her not to be
alarmed, adding these words, " The affair will be carried through.' She insisted on being allowed to keep
the letters and documents, and said that she was
anxious to peruse them all and thoroughly understand
the whole question.'

" The Emperor, who has also transacted business
with M. Thouvenel, in the absence of M. Drouyn de
Lhuys, spoke in very favourable terms on the subject.
He instructed M. Thouvenel to write to Count Walewski in London, so that he might explain the matter
to the English cabinet, and give them to know how
much he was interested in it, with the intention of
himself discussing the subject in higher quarters
when he goes to England with the Empress, as he
will do shortly, on a visit to the Queen.

"A letter from Constantinople which I have just received is to the following effect:—

"'M. Benedetti has been shown the correspondence between Kiamil Pasha and his brother-in-law the Viceroy, and he has succeeded in obtaining the notes written by Reschid Pasha himself, and which served as data for this correspondence. He first of all sought for a direct explanation from the Grand Vizier, whom he charged with having inspired the letter in which the Emperor's name is made such an improper use of. The Grand Vizier defended himself as best he could—that is to say, very badly—laying all the blame upon Kiamil. Instructions were asked for from Paris, the Sultan was made acquainted with what had occurred, and, after several ministerial councils had been held, Reschid Pasha has been dismissed out of deference to the just susceptibilities of the French Government.'

"The upshot of all this is that we shall be left in peace so far as Constantinople is concerned for some time to come. I am therefore free to return to France to carry on my propaganda, and to act in accordance with the programme to which your Highness has been pleased to agree."

To Count Th. de Lesseps, Paris.

"ALEXANDRIA, *May* 19, 1855.

"Before embarking for France I have had a conversation with Mr. Bruce, the British Consul. He himself opened the subject of the Suez Canal, and told me that he had not received a line from his Government, which had not even acknowledged receipt of the documents I had submitted to him. He expressed a hope that my approaching visit to Paris and London would contribute to bring about an understanding, if necessary, between the two Governments, especially since the Viceroy had determined to carry the railway as far as Suez, as this step, very favourably received by public opinion in England, removed, he thought, all pretext for opposing the scheme of a canal. Our conversation took place in presence of Lord Haddo, the Earl of Aberdeen's eldest son. I am much afraid that if Mr. Bruce is sincere in what he says he will not long be the representative of the English cabinet in Egypt.

" I shall find in M. Walewski, our new Minister of Foreign Affairs, a very cordial partisan of the Suez Canal, for in a recent letter from London he promised me his heartiest support for an enterprise of which he had himself spoken to Mehemet Ali when in Egypt fifteen years ago."

Note addressed to Count Walewski, Minister of
Foreign Affairs.

" PARIS, *June* 7, 1855.

" I beg of Count Walewski to be pleased to ask the Emperor for instructions as to my journey to London, whither I am ready to start at once.

" I am of opinion that the wisest plan will be to let the Suez Canal Scheme retain its private character and not allow it to be dependent upon the will of a government which might not be favourable to the project.

" Should I be called upon to reply to any objections or proposals which should be made, may I say that the Imperial Government would· be disposed, concurrently with England, to declare without further delay that at no period should the commercial navigation be interfered with by a belligerent Power ?

" In order not to ruffle foreign susceptibilities, it is essential to point out, whenever the opportunity arises, that the concession of the Suez Canal has not been granted to a Frenchman, or to a French company. M. de Lesseps, as a friend of the Viceroy, has received exclusive powers from him to form a Universal Company, to which *only* the concession will be granted."

Note for the Emperor.

" Paris, *June* 9, 1855.

" In the audience which your Majesty was pleased to accord me you advised me to proceed at once to London, and to get into communication with *The Times*.

" I have the honour to inform your Majesty that I am ready to start, and that Mr. O'Meagher, the correspondent of *The Times* in Paris, with whom I was on very friendly terms during the risings in Barcelona, and when I was Minister at Madrid, was sending to the manager of *The Times* the enclosed letter, which embodies his own views as to the Suez Canal Scheme.

"*The Times*, June 13th, 1855.—' The project for cutting a canal through the Isthmus of Suez begins to occupy so much the attention of the public, notwithstanding the absorbing interest attached to our operations in the Crimea, that it may not be considered out of place to say a few words respecting it. It appears that in November last the Viceroy of Egypt communicated to the Consuls-General accredited to him a firman, in virtue of which M. Ferdinand de Lesseps was authorised to organise a "Universal Company," to which the concession of the construction of the canal should be accorded. The terms of the firman exclude the idea which had been entertained, that the said concession was exclusively granted to a single French subject, or even to a French company. It was granted to an association of shareholders of every country, to be constituted by the person already named as the representative or negotiator of the Viceroy. M. de Lesseps at the outset gave complete explanations to Mr. Bruce, her Britannic Majesty's diplomatic agent and Consul-General in Egypt. In the letter addressed from Cairo, on the 27th of November, to that gentleman the negotiator expressed his earnest desire to avoid everything which could

give the slightest umbrage to national jealousy of any
kind. It is, moreover, affirmed that he contributed in
no slight degree to the completion of the railroad
from Cairo to Suez, the consequences of which have
been so beneficial. M. de Lesseps proceeded to Constantinople, and placed himself in friendly communication with Lord Stratford de Redcliffe, and, through
the Grand Vizier, delivered to him from the Viceroy
of Egypt an official letter, in which the construction
of the canal was described as most useful. He abstained from pressing for the ratification of the Sultan
the moment he perceived a shadow of opposition on
the part of the English Ambassador.' "

"PARIS, *June* 14, 1855.

" M. Thouvenel was yesterday received in audience
by the Emperor prior to his departure for Constantinople, where he has been appointed ambassador, and
the Emperor, in handing him his written instructions,
verbally told him to lose no time in informing the
Porte and the Sultan that it was his wish that the
ratification should be sent direct to the Viceroy,
and in expressing his dissatisfaction should Lord
Stratford de Redcliffe's efforts to prevent this being
done succeed.

" The Emperor's private secretary has informed me
that I shall be able to leave for London towards the
end of the week, so you see that everything is pro-
gressing very favourably. M. Barthélemy St. Hilaire

has begun to lend me his aid. I enclose you a note
which I have received from Baron James de Rothschild, to whom I was enabled to render some little
service while I was Minister at Madrid. I have had
the interview with him which in his note he proposes.
He asked me what my intentions were with regard to
the financial organization of the scheme. I told him
frankly that I did not intend to enter into any positive engagement, that the matter was under careful
consideration, and that I did not intend to bring it
forward until all uncertainty with regard to the execution of the scheme had been cleared up, but that as
soon as ever circumstances allowed he would be one
of the first persons whose co-operation I should seek.
He said that he thought I had chosen a very wise
course, and offered to assist me in any way that lay
in his power. Hearing that I was going to England,
he gave me the following letter to his London house,
which, considering how great is the perspicacity of
this prince of finance, I regard as a very favourable
symptom.

" ' We have the pleasure to introduce to you M.
Ferdinand de Lesseps, who has just arrived from
Egypt, where he has, as you know, been busily engaged in studying the question of making a canal
through the Isthmus of Suez.

" ' We do not doubt that you will be very pleased
to see M. de Lesseps, who proposes to discuss this
subject with you. We beg to commend him to you

most favourably and request you to give the utmost
attention to his interesting communications, the importance of which will be as apparent to you as it is
to us.'

"Upon the other hand, M. Thiers informed me this
morning that Lord Ashburton, a partner in the great
banking firm of Baring Brothers, who is now in Paris,
had written to his firm as well as to his friends in the
most favourable terms of the Suez Canal, and that I
shall be very cordially received by them."

To Count Th. de Lesseps, Paris.

"London, *June* 25, 1855.

"I am in a position to give you a slight sketch of
my first proceedings and of what has come of them.
To begin with, I have had two very long consultations with the principal manager of *The Times.* He
considers that England has no serious objection to
offer against the proposed canal, that those hitherto
raised rest on no solid basis, and that as the article
sent from Paris by Mr. O'Meagher, which appeared
in *The Times* of the 13th, presented the question in a
very clear and favourable light, we might agree to
choose the time most suitable for recurring to the
matter. In the meanwhile he has promised me—and
that is the essential point—not to take part against
the project, and as several letters from an English
correspondent at Alexandria, written in a spirit hostile
to the scheme, had been sent to the paper, this corre-

spondent is to be communicated with and asked to examine the question without prejudice, and ascertain if it is really believed in Egypt that the project is practicable.

"Then again Mr. Reeve, one of the Secretaries of the Queen's Privy Council, who has great influence with *The Times,* to which he is a frequent contributor, though he does not admit the soft impeachment, was very explicit. I was specially recommended to him by M. Barthélemy St. Hilaire, his intimate friend, in whom he has the utmost confidence. He told me that I might be certain of not encountering any preconceived opposition, and he added—

"'It would be degrading that England should have an interest in rejecting a scheme which would be beneficial to the whole world. Upon the contrary, we should derive more benefit from it than anyone else. All that you have to do is to show the public that it is feasible ; that English capital, as well as that of other nations, will be allowed to share in it ; and that there will be no special privileges for any one nation.'

"By the advice of Lady Tankerville, a friend of Lady Palmerston, I called upon Lord Palmerston one morning, with a letter of introduction from Paris. He received me at once, but I thought that I could see at once that his mind was made up on the subject. I entered upon it at once, and asked him if he could spare the time to discuss it with me openly, and not scruple to tell me what his objections really were. He

repeated, word for word, the remarks contained in
Lord Clarendon's despatch to Lord Cowley, which had
evidently been dictated by him, or at all events drawn
up under his inspiration. The subject was a familiar
one to me, so I was at no loss to reply more in detail
than I could to the note which was shown me in Paris
the day before I left. I could not hope in a first conversation, prolonged though it was, to modify or shake
the conviction of a man of Lord Palmerston's character,
but I was pleased to find that my arguments were
unanswerable; that, despite his facility of speech and
lucidity of intellect, he had no serious reply to make.
He evidently had in reserve other objections which
had not yet been produced. With an air of *bonhomie,*
he went on to say :—' I do not hesitate to tell you
what my apprehensions are. They consist in the first
place of the fear of seeing the commercial and maritime relations of Great Britain upset by the opening
of a new route which, in being open to the navigation
of all nations, will deprive us of the advantages which
we at present possess. I will confess to you also that
I look with apprehension to the uncertainty of the
future as regards France—a future which any statesman is bound to consider from the darkest side, un-
bounded as is our confidence in the loyalty and sincerity of the Emperor; but after he has gone things
may alter.'

" I then asked Lord Palmerston to examine at his
leisure all the questions relating to the political side

of the affair, with the conviction that from an impartial
and unprejudiced consideration of them, it would be
clear to him :

" 1st. That England was more interested than any
other nation in the route to India being shortened by
more than three thousand leagues.

" 2nd. That if in the remote probability of its ever
unfortunately happening that France and England
should be embroiled, it would be easy to prove that
the opening of the Suez Canal would not be a cause
of weakness to Great Britain, mistress as she is of all
the important passes and maritime stations between
the metropolis and India.   It must also be borne in
mind that, since the introduction of steam, the conditions of a war between the two countries are different,
and that the French, a people who do not travel much,
would not attack England in India when she was
within two hours of their coast.   I added, moreover,
that if at some future time the execution of the canal
was deemed possible by engineering science, and if
the free capital of all nations saw therein a source
of material profit, irrespective of all political influence,
the Governments of France and England were upon
sufficiently intimate terms to agree upon such measures as would guarantee their mutual interests ; that
I had no mission to treat of this subject, and that my
sole object in coming to London, as delegate of the
Viceroy, had been to ascertain for myself the state of
public opinion on the Suez Canal, and to endeavour

to give in all good faith all the information in my power, both as to the possibility and advantages of the undertaking, as well as to the universal principle of satisfying all the interests which were entitled to consideration in the matter.

" This first conversation was only preliminary. It was very deferential on my part, and conducted with much courtesy by Lord Palmerston, who gave me more of his time than I could have expected that he would. A few hours later I received an invitation from Lady Palmerston to spend an evening at her house.

" I have not yet spoken to you of M. de Persigny, our ambassador, but I may tell you that we seem likely to work in complete harmony with each other, and that I met him the same evening at Lady Palmerston's. Yesterday I dined with Mr. James Wilson, Secretary to the Treasury, a very distinguished economist, who has offered me his services, as has Mr. Edward Ellice, a friend of M. Thiers, and one of the most influential members of the House of Commons. Mr. Ellice has introduced me to several of his colleagues, and he has asked me to dine with him on the 2nd of July, and meet several political and financial personages who will be of use to me. His handsome residence is close to where I am staying, and I have a general invitation to go and breakfast with him in his study at nine every morning. I have had an interesting conversation with M. de Pannizzi, the librarian of the British

Museum, with reference to a work in English which I am preparing, and which it is generally thought will be of considerable use."

To Baron de Bruck, Minister of Finance at Vienna.

"London, *June* 28, 1855.

"My journey to Vienna will be somewhat delayed by my prolonged stay in London, where I have to meet difficulties and objections arising from a mistaken appreciation of the affair. I must not fail to tell you that the reports and speeches of Mr. Robert Stephenson, a member of the company formed for studying the question in 1847, and the recent article in the *Revue des Deux Mondes*, in favour of a roundabout route, by means of a bridge canal carried along the two branches of the stream and fed by the water of that river, have contributed in no small degree to create erroneous impressions in the public mind, and to make some people believe that the piercing of the isthmus is an impossibility, and to make others believe that it can only be executed at a cost of labour and money out of all proportion to the revenue to be derived from it.

"The Viceroy still defrays all the expenses of the preliminary experiments, and it will not be until the commission of engineers, well versed in hydraulics, to which he proposes to entrust the drawing up of the definite scheme, has reported upon the matter, that we shall proceed to organise the company and open a subscription for shares."

To Count Th. de Lesseps, Paris.

"LONDON, *June* 30, 1855.

"I yesterday had an interview with Lord Clarendon, the Secretary of State for Foreign Affairs, who
certainly has no preconceived hostility to the project,
like Lord Palmerston. I think that the best thing I
can do is to give you the substance of the conversation which took place, by putting it in the form of a
dialogue, though I may add that we began by having
a chat over our friends in Spain, where we first met
in 1848.

"*Myself.*—Entrusted by the Viceroy with the preparations for organising a Universal Company for the
piercing of the Isthmus of Suez, I was desirous of ascer-
taining for myself the state of public opinion in England, and of explaining to any one who was anxious
to be enlightened on the subject:—1st, That the
affair has not been undertaken by any government, or
to the exclusive profit of any nation. 2nd, That the
enterprise is materially practicable—that is to say,
that the estimated expenses will be proportional to the
profits accruing from the traffic. 3rd, That there is
no intention of soliciting the intervention of the
British Government, or of making at present any
appeal to investors. 4th, That the most able of
European engineers will be called upon to decide as to
the possibility of the work being carried out, and as
to its cost. 5th. That when once the enterprise has
been found to be practicable, investors, large or small,

T 2

will be at liberty to subscribe without any regard to politics. 6th, That the Viceroy, having completed the railway from Alexandria to Cairo, at the cost of the Egyptian treasury, and being now engaged in making the final section from Cairo to Suez, had been anxious to give full satisfaction to England. And 7th, That the Suez Canal, having been of his own free will made over to private enterprise, there was no fear of the resources of the country which he ruled being imperilled, and that his only aim was to further the interests of Egypt and of his Suzerain. All that I now ask you is to examine the question calmly and impartially, being convinced that a mind so enlightened as yours will not admit it to be possible that an event so profitable to the moral and material interests of the whole world can be detrimental to the power or commercial relations of England.

"*Lord C.*—I will not conceal from you that the tradition of our Government has, up to the present, been hostile to the making of a canal through the Isthmus of Suez. I have myself, since I have had to deal with this question, been compelled to conform my opinion accordingly, and I confess it is not favourable to the scheme.

"We then discussed the objections raised in Lord Cowley's note.

"*Myself.*—It seems after what has been said by yourself as well as by me that the opinion which you may have formed before you were acquainted with the new aspect of affairs is open to modification or, at all events, that, if you admit that the subject is one deserving examination, your opinion must be based upon reasons and arguments of a higher kind. It is easy to understand that a time when the two governments of France and England ordered their agents to oppose one the railway and the other the canal, each of them should hold to its opinion, however unreasonable it may have been. There was in that case a *parti pris* on either side. But a great change has taken place since then. The intimacy and the sincerity of the alliance between the two countries does not admit of this antagonism existing, especially in matters of progress and of general interest which are beneficial to the whole world. Consequently, the French agents, far from running counter to the English agents in regard to the Egyptian railway, have, as you must know, cordially supported them. Surely England will not cling to the remnants of an antagonism which has been loyally and entirely foresworn by France. The sentiments of the members of the English Cabinet are too well known for us to doubt what their decision will be. Therefore, as I repeat, all that I ask of you is to give an impartial consideration to the affair.

"*Lord C.*—I am much obliged to you for speaking to me so frankly, and what you say deserves to be taken into careful consideration. You may rely upon my doing as you wish, and examining the question deliberately, without the slightest prejudice.

"Lord Clarendon then went on to speak of Constantinople, and said that Lord Stratford de Redcliffe complained very much of Benedetti, to which I replied that this was a case of the wolf complaining of the lamb. As he told me that Lord Stratford de Redcliffe had not originally taken any action against the project at first, I told him how Reschid Pasha himself had confided to me the difficult position he was placed in, owing to the active steps taken by Lord Stratford de Redcliffe on the one hand, and the passive attitude of Benedetti on the other.

"'I may tell you,' added Lord Clarendon, 'that Lord Stratford de Redcliffe speaks in the highest terms of the excellent personal relations he has had with you. To revert to the present relations of the French and English cabinets, I can assure you that my colleagues and myself consider ourselves members of one and the same Cabinet. Our confidence in the Emperor and Count Walewski, whose loyalty has done much to tighten the bonds between the two countries, is complete and unrestricted.'

"I am pleased to find that Lord Clarendon, in speaking thus, is expressing the unanimous opinion of all classes in this country.

"The only fear seems to be lest the very sincere desire which exists for an alliance should not be shared by public opinion in France, and fashionable people who have been in France, and passed some time in Paris society, help to accredit this hesitation. This is the real cause of the mistrust with which so many English politicians regard the future. I assure them that they are mistaken, that the alliance of the two countries is quite as much a national alliance in France as it is in England, that the ancient party feelings have ceased to find any echo, and that the hesitations or doubts produced in English policy through the mistrust of the future can only serve to give arms to the adversaries of the alliance, and eventually, perhaps, to deprive it of its national character."

To His Majesty the Emperor, Paris.

"LONDON, *July* 4, 1855.

"The interest which your Majesty has deigned to take in the great enterprise for opening the Isthmus of Suez, has emboldened me to lay before you the result of my preliminary steps in London.

"The Queen's Ministers have shown a disposition to examine the question carefully. They have made a point of declaring that their objections were raised in good faith, and without any feelings of mistrust towards your Majesty's Government. The editors of the *Times* and other newspapers have assured me that they were well disposed. I have met with sympathy, promises of support, and even active assistance, from a great many men of influence in politics, science, industry, and commerce. Among them I may mention Lord Holland, that old and tried friend of France ;

the Duke of Northumberland, Lord Seymour, Mr. Edward Ellice, M.P.; Sir Richard Gardner, M.P.; Mr. Rendel, the leading hydraulic engineer in England; Mr. Charles Manby, secretary of the Institute of Civil Engineers; Mr. Reeve, secretary of the Queen's Privy Council; Mr. James Wilson, secretary of the Treasury; Mr. Morris, manager of *The Times;* Mr. Oliphant, one of the managers of the East India Company; Mr. James Welch, captain in the Royal Engineers, secretary of the Admiralty, and author of a treatise on 'The Advantages of the Suez Canal from a British Point of View;' Mr. Pannizzi, librarian of the British Museum; Mr. Thomas Hanley, governor of the Bank of England; Mr. Powles, secretary of the Dock Company; Messrs. Anderson, Wilcox, and de Zuluela, directors and founders of the P. and O. Steam Company; Sir W. G. Ouseley, minister plenipotentiary, Mr. Thomas Wilson, author of the project for a 'Canal from the Danube to the Black Sea;' and the chiefs of the foreign embassies and legations.

"None of the enlightened men with whom I have discussed the question have been prepared to say that an event which would be profitable to the interests of the whole world could be injurious to the power or the commercial relations of England. They dismiss all idea of a preconceived hostility to the scheme; upon the contrary, they assert that, if it is practicable, their country has everything to gain by it, and they would be very sorry for it to be supposed in France that what would be beneficial to other nations would not be equally so to England.

"In fine, I have acquired the conviction that the enterprise of the Suez Canal, far from troubling in the smallest degree the relations of France and England, will contribute, upon the contrary, after the exchange of frank and open explanations, to bring out in a very clear light the sincerity of the alliance between the two countries.

"The favour with which the question was received by public opinion, the publications which are being prepared, the influence of the interests of trade and navigation, and the desire to give a mark of confidence in your Majesty, cannot fail to bring over those members of the English Cabinet whose opposition might, a short time ago, have justified the idea of an energetic resistance on their part, which there seems no longer any reason to apprehend."

THE QUESTION OF THE ISTHMUS OF SUEZ SUBMITTED
TO ENGLISH PUBLIC OPINION.

"LONDON, *July*, 1855.

"*Aperire terram gentibus.*

"In October, 1854, I left Europe for Egypt, upon the invitation of the Viceroy, Mohammed Said, with whose friendship I had been honoured for twenty years. I had no mission from my Government, and it was in the course of a journey with the Prince, from Alexandria to Cairo, across the Libyan Desert, that the question of piercing the Isthmus of Suez was first discussed between us.*

" I have come to England to place the matter clearly before the eyes of the public. I appeal to the interests and am content to rely upon the judgment formed by the East India Company,. the traders with Australia, Singapore, Madras, Calcutta, and Bombay, the merchants of the city, the shipowners of London and Liverpool, the manufacturers of Manchester, the ironmasters, the makers of machinery, the P. and O. Steam Company, the managers of banks and other large businesses, the commercial associations, and the owners of coal mines who in 1854 exported nearly four and a half million tons of coal, representing a value of £2,147,156, and who, by the opening of the Suez Canal, would find these enormous exports considerably increased.

" It has been objected that the Turkish Government ought to concern itself about this project; but as in every question where the principles are just, the foreseen consequences are inevitable. No matter from what point the enterprise of the Suez Canal is considered, it will be found to be of advantage to all the world.

" Turkey can only shake off its present state of torpor by obtaining from Europe capital and intelli-

*Note of the Translator.—The greater part of this letter describes the preliminary steps taken by the Viceroy and M. de Lesseps to ascertain the nature of the work and its cost, the account of which will be found in the preceding pages.

gence. The prosperity of the East is now dependent upon the interests of civilization at large, and the best means of contributing to its welfare, as well as to that of humanity, is to break down the barriers which still divide men, races, and nations."

Circular to the Members of Parliament, Merchants, Indian Shipowners, &c.

"LONDON, *August* 8, 1855.

" I have the honour to send you a copy of my work relating to the piercing of the Isthmus of Suez. I hope that, after having perused the various documents submitted to your notice, you will favour me with your views as to the advantages of this important undertaking, in the successful accomplishment of which I believe Great Britain to be more interested than all the other nations.

" The alliance which exists between our two countries induces me to attach great importance to the views of the most enlightened Englishmen concording with those which prevail in France on this subject. You will observe that the project of the engineers of the Viceroy of Egypt is to be submitted, before being carried into effect, to a Commission selected from among the most celebrated engineers in Europe. Mr. Rendel, well known for the remarkable works executed by him in English ports, will be a member of this Commission.

" I shall be obliged if you will address your reply

either to my Paris residence, 9, Rue Richepance, or else to Messrs. Baring Brothers, or Messrs. Rothschild in London.*

"The Commission spent upwards of six weeks in Egypt, and after examining carefully the Isthmus of Suez and the various plans proposed for tracing the canal, submitted the following summary report to the Viceroy:—

"ALEXANDRIA, *January* 2, 1856.

"Your Highness summoned us to Egypt to examine the question of the piercing of the Isthmus of Suez. While supplying us with the means of deciding, *de. visu*, as to the merits of the different solutions proposed, you requested us to lay before you the one which was the easiest, the safest, and the most advantageous for European commerce. Our explory, favoured by magnificent weather, facilitated and shortened by the ample material means placed at our disposal, is completed. It has revealed to us innumerable obstacles, not to say impossibilities, for taking

* Note of the Translator.—It may be added that this scientific commission was not finally completed until the beginning of October, among the members appointed being, in addition to MM. Renaud and Lieussou, for France, M. de Negrelli, for Austria; Mr. Rendel, for England; Herr Conrad, Inspector of the Waterstat, and President of the Society of Civil Engineers, for Holland; and the Privy Chancellor Lentze, for Prussia. MM. Linant Bey and Mougel Bey came over from Egypt to meet them in Paris, the first meeting being held at M. de Lesseps's residence on the 30th of October, and left Marseilles for Alexandria on the 8th of November.

the canal by way of Alexandria, and unexpected facilities for establishing a port in the Gulf of Pelusium.

"The direct canal from Suez to the Gulf of Pelusium is, therefore, the sole solution of the problem for joining the Red Sea to the Mediterranean; the execution of the work is easy and the success assured. The results will be of immense importance for the trade of the world. We are unanimous in our conviction upon this point, and we will develop our reasons for it in a detailed report, reinforced by tydrographical charts of the Bays of Suez and Pelusium, outlines showing the relief of the soil, and borings indicating the nature of the soil through which the canal will pass. This is a long and minute work which will occupy several months; but in the meanwhile we beg to acquaint your Highness with our conclusions, which are as follows :—

" 1st. The route by Alexandria is inadmissible, both from the technical and economical point of view.

" 2nd. The direct route offers every facility for the execution of the canal itself, with a branch to the Nile, and presents no more than the ordinary difficulties for the creation of two ports.

" 3rd. The port of Suez will open on to a safe and large roadstead, accessible in all weathers, and with a depth of about thirty feet of water within a mile of the shore.

" 4th. The port of Pelusium, which according to the

draft scheme was to be at the extremity of the Gulf, will be established about seventeen miles further to the west, at a point where there are 25 feet of water within a mile and a-half of the shore, where the anchorage is good and getting under way easy.

" 5th. The cost of the canal, and of the works connected with it, will not exceed the figure of £8,000,000 given in the draft scheme of your Highness's engineers.

" The members of the International Commission of the Suez Canal.

(Signed)   CONRAD, *President.*

A. RENAUD, DE NEGRELLI, McLEAN,*

LIEUSSOU, *Reporter and Secretary.*

" A copy of this report was sent to the following supporters of the enterprise from Alexandria on the 4th of January : —

"Jomard, member of the Institute (Paris); Morris, *The Times* (London); Thouvenel (Constantinople); Brusi (Barcelona); Erlanger (Frankfort); Couturier (Marseilles); Charles Manby (London); Theodore Pichon (Smyrna); Edmond de Lesseps (Beyrout); Revoltella (Trieste); Flury-Hérard (Paris); Count Th. de Lesseps (Paris); Count Walewski (Paris); Damas-Hinard (Paris); De Chancel (Paris); Marcotte (Marseilles);

* Note of the Translator.—Mr. McLean was the English Commissioner in place of Mr. Rendel, whom illness prevented him going out to Egypt.

Senior (London); Ellice (London); James Wilson (London); Thiers (Paris); Archduke Maximilian (Vienna); Baron de Bruck (Vienna); H.R.H. the Duke de Brabant (Brussels); Lord Holland (London).

"As the principal opposition to the scheme still came from the English Government, I determined to pay a second visit to London, and while in Paris, on my way from Egypt, I had a long conversation with Lord Clarendon, at the close of which, after hearing all I had to say, he held out the hope of our being able to come to an understanding, adding, 'As you are going to London, please repeat to Lord Palmerston the substance of our conversation. We can discuss the matter together again, for we shall meet in London in a few days.'"

To M. de Negrelli, Vienna.

"LONDON, *April* 17, 1856.

"I enclose you the text of an important conversation I had with Lord Clarendon just before I left Paris, and this will explain to you the motives of my journeying to England. Having been here a week, I have not yet had an opportunity of discussing matters with Lord Palmerston, owing to the sudden death of his stepson, Lord Cowper. But I have not been idle, and I am carefully preparing to form an English committee composed of eminent men who will render the same success which you and Baron de Bruck do in Austria.

" It appears from all I can gather that there is not, as I had foreseen, much help to be expected from official diplomacy, and that we must rather look to the accomplishment of facts which in due course will receive the sanction of diplomatists, because it will then be their interest to concern themselves with what has been done.   In the meanwhile you may rest assured that it would be dangerous rather than otherwise for a spontaneous diplomatic intervention, which might have the effect of alarming the Viceroy, and lowering his situation in Egypt under the pretence of taking guarantees against him.   There was already some talk of taking these guarantees at the conferences which have been held, upon the ground that the making of the canal would increase his power.   Therein, I repeat, resides, to my mind, the most serious difficulty which we have to foresee and take into serious consideration. As matters stand, the object which we have to keep in view is to induce the European Governments, that of England in particular, to place no obstacle in the way of the ratification which the Viceroy has asked for from the Sultan, and which the latter is disposed to grant him.   This once obtained, we are masters of the situation, and we avoid the danger which I have pointed out to you.   Talk the matter well over with Prince Metternich and Baron de Bruck.   You may be certain that if the Viceroy saw that any Power had the least idea of lowering his regular authority, he would give up all idea of getting Europe to assist in the making of the canal. I, for my part, am too sincerely his friend not to follow him in this course. Our basis and our principal supports are in Egypt. If in the course of the last fifteen months I had looked elsewhere for support, I should have done nothing, and matters would not have reached the point they have.

"Whatever progress the matter has made, it only remains with the Viceroy to•prevent it being carried any further. I need not, therefore, dwell further upon this subject, the full gravity of which you who know the character of Mohammed Said, and who are so well acquainted with Egypt, will at once understand.

"I break off this letter to go and meet Lord Palmerston. While confirming what Lord Clarendon said to me in Paris, he persists in his opposition, and did not make any secret of the fact that Lord Stratford will continue to oppose us, not now in the interests of England, but in the alleged interest of the Ottoman Empire. This tactic shows that the enemy of the canal is driven to his last retrenchment, and I am going to prepare my parallels and pursue with prudence, but with more perseverance and vigour than before, my appeal to public opinion in England. One campaign the more will not discourage me, and in the meanwhile the matter will ripen and assume a consistent shape, which will add to our force.

"Mr. McLean is fully convinced as to the success of the enterprise. I have arranged with him and with Mr. Rendel that, in accordance with the wish expressed

by all the other members of the International Commission, the general meeting should be held in Paris."

To M. Barthélemy St. Hilaire, Paris.

"LONDON, *April* 7, 1856.

"I found Lord Palmerston just what he was in
1840, full of mistrust and prejudices with regard to
France and Egypt. He was very polite, and was in
some respects very frank, but after hearing me read
the *résumé* of my conversation with Lord Clarendon,
he spoke to me, with regard to the Suez Canal, in the
most contradictory, the most incoherent, and, I will
even add, the most senseless fashion imaginable. He
is firmly convinced that France has long pursued a
most machiavelian policy in Egypt against England,
and that the fortifications of Alexandria were paid for
by Louis Philippe or his Government. He sees in the
Suez Canal the consequences of this policy. Upon
the other hand, he persists in maintaining that the
execution of the canal is materially impossible, and
that he knows more about it than all the engineers in
Europe, whose opinions will not alter his. Then, re-
gardless of the fact that he had just proved the scheme
to be impracticable, he indulged in a long tirade
upon the drawbacks which would result for Turkey,
and for Egypt herself, from the Viceroy's concession
and the realisation of the enterprise. Finally, he
declared that he should continue to be my adversary
without any sort of reticence. I could not help ask-

ing myself now and again whether I was in the presence of a maniac or a statesman. There was not one
of his arguments which would hold water for five
minutes in a serious discussion. I replied to all his
objections as best I could, but I saw that it was only
a waste of time to prolong the discussion. Seeking
as I do to have the ground clear, I am not at all sorry
to know how things stand, and I shall prepare my
batteries accordingly.

"Please report all this to M. Thiers, and let me
know what he thinks. I should not be surprised if
Lord Palmerston, his old opponent in 1840, believed
him to be the author and the continuer of the machiavelian policy in question."

To M. Ruyssenaers, Alexandria.
"LONDON, *April* 21, 1856.

" We now know the true motives of Lord Palmerston's opposition. It is that he is afraid of favouring
the development of Egypt's power and prosperity.
Fortunately, this is not the kind of motive likely to
discourage the Viceroy in the pursuit of his noble
enterprise.

" I suspected this to be the case some time ago,
and I pointed it out to his Highness last year when
speaking to him of a despatch of the ex-Governor of
India, in which he said, that if England should ever
succeed in obtaining a footing in Egypt, as she had
done in India, she would be the mistress of the world.

Nor will his Highness forget that in a document of 1840, published by Mr. Urquhart, First Secretary of the British Embassy at Constantinople, the English Ambassador, Lord Ponsonby, wrote to the Grand Vizier, that the policy of England and the Porte should be to drive Mehemet Ali and the whole of his family naked into the desert. There can be no doubt that although this idea, now impossible of realisation, is only now to be found in a few very wooden old heads, it is desirable to guard against the irritation which the approaching execution of an enterprise destined to have the very opposite effect will provoke. As long as a partisan of the policy which consists in weakening Egypt is at the head of the English Ministry, it is necessary to paralyze its evil intentions by acting with extreme prudence, by continuing to enlighten public opinion, by marching prudently and without undue hurry towards the achievement of the fact. The Viceroy will see in my advice and my conduct the best proof of my desire not to compromise him. If I thought more of the canal than of him, nothing would be easier for me than to make the affair over to large capitalists, who would carry it through much more rapidly; but I am determined that he shall remain the master of it, and that it shall serve to consolidate and fortify his political situation.

"Even if there had been no canal scheme, the Viceroy may rest assured that there are certain Englishmen who would have found some excuse for attacking him in the same way, and soon it will be the canal which will act as a lightning conductor for him.

" Since I have been in England I have been constantly rectifying the erroneous ideas which are kept alive about Egypt, and which are for the most part propagated by certain ill-disposed journals.

"Mr. McLean entertained me the day before yesterday at a somewhat important banquet given at the Trafalgar Hotel, Greenwich. He had got together about thirty guests, included among them being the most celebrated engineers in England, manufacturers, merchants, and bankers. Mr. McLean, in proposing my health, spoke of the hospitable way in which he and his colleagues of the International Commission had been received in Egypt, and expressed, upon his own behalf, as upon behalf of his friends, the hopes that England entertained for the realisation of the scheme and for the success of my efforts. This toast and my reply to it were received with loud applause. My object is to bring public opinion in England to pronounce in favour of the Isthmus of Suez, so that the English Government may be led to follow the same policy as France.

"A gentleman named Wyld, geographer to the Queen, formerly a member of Parliament and owner of ' The Great Globe,' Leicester Square, gives ocular demonstration, three times a day, at this latter establishment, of the advantages which navigation will derive from passing through the Isthmus of Suez instead of going round by the Cape. He is now having made a large relief plan of the same dimensions as that of Sebastopol, which attracted a great many people. This popular mode of propaganda is excellent. In my conference with Lord Palmerston, described in my letter to M. B. St. Hilaire, the Prime Minister admitted that the English Ambassador at Constantinople had maintained that the Viceroy of Egypt did not require the authorisation of the Sultan for the railway from Alexandria to Suez, but that the situation was different in respect to the canal. My reply was that I saw no difference except that the English Government wanted the railway, and does not want the canal.

"In short, tell his Highness that this ill-will will eventually be paralyzed, and that with his perseverance and continued help, the obstacles and the difficulties encountered will only serve to aggrandise his position and render the success more complete."

To M. Thouvenel, French Ambassador at Constantinople.

"LONDON, *April* 22, 1856.

"The following information, upon the exactitude of which you may rely, will interest you, and may perhaps be of some use to you.

"After the banquet given by the Emperor to the Plenipotentiaries of the Paris Congress, Aali Pasha, the Turkish Ambassador, came up to His Majesty and asked him what he thought of the Suez Canal ques-

tion. He added that his master attached very great
importance to it in every way, but that he was
anxious to know what the Emperor's views were.
Napoleon replied that he took the greatest possible
interest in this scheme, which seemed to him beneficial for everyone, that he had studied it in all
its aspects, that he had made himself acquainted with
all the documents bearing on it, and that he wished
it every success; that the enterprise, noble a one as
it was, had given rise to certain resistances and
objections, especially in England; that for his part
he could not see that these objections were founded,
and that he hoped to see them removed; but that at
the same time he would not hurry on things, for fear
of compromising their success, and that, relying upon
the happy alliance which united the two peoples, he
looked to the future, and to a very near future, for an
agreement upon this question. Aali Pasha said that
his master would be very pleased to hear of the
sympathies expressed by the Emperor, and that he
was himself favourable to the affair, despite certain
divergencies upon secondary points and certain pre-
cautions to be taken in the interests of the Suzerainty
of the Porte; but, irrespective of these objections on
points of detail, the Porte none the less looks with
favour upon this great work, which will be so profitable to Egypt, and in which she also hopes to have
her share of profit.

"The Emperor, who seemed to acquiesce in all Aali

Pasha said, then moved away and called upon Lord Clarendon, and asked him what he thought of the Suez Canal, relating to him what Aali Pasha had just said, and what he had said in reply. Lord Clarendon, somewhat taken by surprise, replied that the affair was a very important one, that he had not thought over it sufficiently to give an off-hand reply, that he must refer to his Cabinet, and that, moreover, the execution of the scheme was impossible. The Emperor, while admitting that the affair was one demanding reflection, maintained that the execution was possible, and that science had pronounced definitely on that score. As Lord Clarendon held to his view, the Emperor said that, admitting the execution of the scheme to be practicable, and reasoning upon this hypothesis, what was England's view? Lord Clarendon then declared that, from the point of view of English trade, there could be no objection, and that England would benefit considerably, but that as regarded the relations of Egypt and Turkey, the matter was a very delicate one, and that the Viceroy had no right to make the canal without the authorisation of the Porte. The Emperor reminded him of the favourable dispositions of the Porte, and there the conversation ended. I must say that it seems to me to be decisive of the matter, and the conclusions to be drawn from it are as under :—

" 1st. To treat with the utmost deference the susceptibilities of the Porte, ascertain precisely what are the objections of *detail*, in which, of course, will be traced the inspirations of Lord Stratford de Redcliffe.

" 2nd. To show equal deference for the susceptibilities of the Viceroy, whom the adversaries of the canal would be glad to bring into conflict with his Suzerain.

" 3rd. The opinion expressed by the Emperor will carry great weight with the Porte and even in England.

" 4th. According to this declaration of Lord Clarendon, English public opinion must be appealed to and English interests engaged in the enterprise."

To M. Ruyssenaers, Alexandria.

"Paris, *May* 6, 1856.

" I prolonged my stay in London and only arrived here two days ago. My campaign in England will bear fruit. I have formed some very excellent acquaintances. I was presented to the Queen, and I also had a very long conversation with Prince Albert, who took me to his study and got me to inform him exactly of what the projected works on the canal were. He told me that the Duc de Brabant, who was interested in the enterprise, had already recommended it to him. I was received in the kindest way possible by the Duke of Cambridge, who expressed to me very freely, and without the slightest reserve, his sympathies with the project. Moreover, I have availed myself of every possible opportunity for saying what I thought, so that no confidence should

be placed in the systematic vilification of the Viceroy,
in which certain journals have recently indulged. I
have quoted positive facts which show the situation
in its true light, and allow of Mohammed Said being
judged as he deserves to be judged, notwithstanding
errors difficult to avoid in a country the administration of which is not yet completely in working order.
I have been treated to a very significant demonstration from the Geographical Society of London, which,
as you know, is composed of very influential men.
First of all, I was invited to dinner by the Society at
their club, Lord Sheffield taking the chair. My
health was drunk in a toast which referred in eulo-
gistic terms to my efforts to bring about the piercing
of the Isthmus of Suez. Mr. Gladstone, a cousin of
the Minister,* then said, speaking in excellent French:
'M. de Lesseps, if in this country we have not been
so prompt as other nations to welcome your enterprise, it is because of our character and habits. But
once we are convinced, we go further and sometimes
show more perseverance than any of our neighbours.
For my own part, I entertained at first considerable
doubts, which are not yet entirely dissipated; but I
am only too anxious to be persuaded, and I heartily
wish you success.'

" I thanked my hosts for their interest, which I

* Note of the Translator.—I leave the responsibility of the statement as to Lord Sheffield and Mr. Gladstone being cousins to
M. de Lesseps.

was glad to find expressed by so distinguished a
company of travellers and savants, in the success of
an enterprise certainly destined to enlarge the domain
of geographical science and facilitate its discoveries.
As I had been told that many of the members present
would not be able to remain for the meeting of the
Society, I entered into some detail with regard to the
explorations of the International Commission and to the
result of its labours. Questions were put to me with
regard to the danger of an accumulation of sand and
the objections urged by the *Edinburgh Review*, and
my replies seemed to satisfy my questioners. I was
then taken to the meeting of the Society, which was
presided over by Mr. (afterwards Sir) Roderick Mur-
chison, and after several speakers had dwelt upon the
importance of opening prompt and easy communications with the various nations of the earth, he called
upon me. The meeting was a very crowded one, and
included a great number of ladies. My rising was
the signal for loud applause, and I was again heartily
cheered at the conclusion of my speech, which, at the
request of the secretary, I afterwards wrote out and
sent to him for publication in the Society's journal.
It was as follows :—

" 'Captain Fitzroy, speaking of a project for making
a canal through the Isthmus of Darien, told you just
now in eloquent terms how the realisation of many
great enterprises which seem almost chimerical till
they come to be studied, becomes apparent to all the

world after they have been carefully examined upon the spot. I hope it will be so for the interoceanic canal in question, and as regards the piercing of the Isthmus of Suez, upon which the chairman has requested me to address you, I can assure you that the enterprise is perfectly feasible.

" ' The majority of the Commission, comprising the most eminent engineers in Europe, which was appointed to study the question, went to Egypt, and was unanimous in declaring that the canalising of the Isthmus of Suez, and the creation of two ports on the Red Sea and the Mediterranean, were easy and certain operations.

" ' The roadstead of Suez is vast and safe. More than five hundred vessels could ride there at the same time. It has a depth of from 16 feet to 42 feet, with a very good bed of mud for anchorage. The English corvette *Zenobia* has been there for three years as a coaling ship for the East India packets; she is stationed in the part most exposed to the wind, and during these three years she has never dragged her anchor, her cables have never been broken, and her communication with land has not once been cut off—which is more than can be said of many very good ports. There are two clean and deep channels, broad enough to wear ship in any weather, and with a depth of from 50 to 65 feet, by which ships can reach the anchoring ground. The Commission has been able to conclude from this that the Suez roadstead has all the qualifications required for forming the head of the canal from sea to sea.

" 'Along the whole course of the isthmus, from Suez to Pelusium, the International Commission encountered no difficulty in the way of digging the canal, nor for keeping it open, the ground being very level and the geological composition of the soil very favourable. The soundings, which the Commission verified, established the fact that the soil of the isthmus is in most places formed of a first stratum of agglutinated sand, of a second stratum of clay, and of a third stratum of calcareous marl, until the plastic clay is reached at a depth of from 36 to 40 feet below the level of the seas.

" 'During our excursion in the isthmus the Viceroy had sent the steam frigate *Le Nil* into the Gulf of Pelusium, where M. Larousse, the engineering hydrographer, had been making numerous soundings and taking a hydrographic chart of the bay. It was found that outside the line of coast there is a zone of fine sand, similar in description to that of the shore, which has a depth of 33 feet, beyond which begins a zone of mud offering excellent anchorage, and extending right out to the deep water of the Mediterranean. The part of the bay in which there is the deepest water is that opposite Tannis, where there is a depth of over 25 feet within a mile and a-half of the shore, along a distance of thirteen miles, from the mouth of Oum-Fareg to that of Gemileh. That is the part selected by the Commission for the Mediterranean entrance of the canal. There is nothing unusual in jetties of from one to two miles long, and at the points which they will occupy there is every facility for vessels anchoring and making sail.

" 'I am about to bring out a pamphlet which will give the text of the Commissioners' report, as well as a reply to the *Edinburgh Review*, which has published some very erroneous information as to the practicability of the enterprise. The errors into which the *Review* has fallen are excusable, because at the time when it treated the question, it had not before it the result of the investigations made by this Commission.

" 'In a country where there are no bounds to the freedom of public discussion, good causes always triumph in the end.

" 'The English publication referred to above is about to appear, and before leaving London I arranged for the formation of a local committee composed of a member of the East India Company, a member of the Bank of England, two financial notabilities in the city, Mr. Powles, Secretary-General of the Docks Company, and two English engineers on the International Commission. But this committee is not to act until the Sultan's ratification has been obtained.

" 'We must now await the meeting of the Commission of Engineers and the movement of opinion in the city before launching the affair upon a fresh phase: the soil being well prepared, we need only arm ourselves with a little patience until the harvest ripens, watching in the meanwhile the seed which we have sown.' "

To His Highness Mohammed Said, Viceroy of Egypt.

"Paris, *May* 20, 1856.

"Although I never fail to let your Highness know by each mail any facts likely to be of interest to you, I cannot refrain from writing to say how much I was touched by your affectionate letter of April 26th, though I had no need of this fresh evidence that I might count upon the continuance of a friendship in which my confidence knows no bounds.

"I had for a long time observed that the adversaries of your Highness were instinctively the adversaries, either open or secret, of the canal. This being so, I was not astonished at the campaign which they instituted as soon as I had left Egypt. But it is sometimes wise to profit by the conduct of one's enemies, and when those who occupy an exalted position are not afraid to hear the truth, when they have sufficient intelligence to examine calmly the attacks levelled against them, and sufficient good faith to be able to distinguish between what is criticism and what is calumny, your enemies, instead of injuring you, have rendered you service. Criticism, even if ill-natured, is to be met by repairing the errors to which it points, while calumny is always to be confounded by the evidence of positive facts and perseverance in conduct free of reproach.

"I can assure your Highness that all the good you have done is appreciated in Europe, and that you must not judge public opinion by the grotesque observations of a few discontented and untruthful persons, who have attempted to form in Egypt a committee for calumniating and denouncing the Government. The truth is making itself more and more apparent every day. Some flatterers might tell your Highness that you need not condescend to justify yourself; but it is always wise to give full publicity to facts, for else the good-for-nothing people, who are as a rule very active, would have things too much their own way with honest people, who are not inclined to put themselves out of the way and do not suspect evil. I have arranged so that the press of all countries shall be kept well informed, and the relations which have been established towards this end are entirely disinterested."

SUMMARY OF THE RESOLUTIONS AGREED TO BY THE INTERNATIONAL SCIENTIFIC COMMISSION SIX SITTINGS HELD ON THE 23RD, 24TH, AND 25TH JUNE.

Transmitted to various Correspondents in all countries, pending the drawing up and publication of the report entrusted to a Special Commission.

"PARIS, *June* 25, 1856.

"1st. The Commission has rejected the system of indirect routes through Egypt, and has adopted the principle of the direct route from Suez to the Mediterranean.

" 2nd. It has rejected the plan for supplying the maritime canal with the Nile water, and has adopted the plan for supplying it with salt water.

" 3rd. It has discussed the advantages and drawbacks of a canal with continuous banks from one sea to another, and has in the end decided that there shall be no banks where the canal passes through the Bitter Lakes.

" 4th. As the interposition of the Bitter Lakes left open will have the effect of deadening the tidal currents, the Commission was of opinion that locks at the two extremities of the canal would not be necessary, subject, however, to the possibility of having to establish them afterwards.

" 5th. The Commission has adhered to the breadth of 325 ft. at the water line and 150 ft. at the bottom for the whole of the distance (12½ miles), which is to be walled in between Suez and the Bitter Lakes, reducing it by one-fifth in the remainder of the canal.

" 6th. The outline of the preliminary project of the engineers appointed by the Viceroy is retained.

" 7th. Entrance from the Mediterranean (Port Said).—The Commission adopts for the port of Port Said the proposal for jetties made by those of its members who went to Egypt, except that the width of the channel shall be 440 yds. instead of 550 yds., and that a second basin shall be made.

" 8th. Port of Suez on the Red Sea.—The Commission accepts the site and direction of the canal, but the breadth to be 330 yds. instead of 440 yds., and a second basin to be made.

" 9th. The Commission is of opinion that the coasts both of Egypt and the Red Sea should be provided with first-class lighthouses against the opening of the canal.

" 10th. A port for revictualling and repairs and graving dock to be formed in Lake Timsah.

" 11th. With regard to the auxiliary canals of sweet water derived from the Nile, the Commission declares that it leaves the selection of the best mode of execution to the judgment of the engineers who may be appointed to superintend the works, by arrangement with the Egyptian Government.

" 12th. Lastly, that it appears from the detailed information given by officers of the Navy who have sat upon the Commission that the navigation of the Red Sea is as good as that of the Mediterranean and the Adriatic, and his opinion, accepted by the Commission, sums up in so many words the judgment of Captain Harris, who has made the voyage from Suez to India twenty times."

To Count Th. de Lesseps, Paris.

" VIENNA, *July* 8, 1856.

" After the International Commission had closed its sittings, it was my duty to go and report to the

Viceroy the definite results of its deliberations. But, at the same time that I submitted to his Highness the scientific consultations of the engineers, I deemed it advisable to let him have the political consultation of the Nestor of European diplomatists.

"Please communicate to Count Walewski the opinion of Prince Metternich, which I committed to writing after my interview with him, and to the accuracy of which he was pleased to testify."

Opinion of Prince Metternich.

" His Highness the Viceroy had the right to decree the making of the Suez Canal. All the measures taken by him merit the assent of the statesmen of Europe ; but in a question of this importance, on which it was to be expected that foreign policy would have something to say, he was well advised in applying for the ratification of the Porte.

" The official approval of an enterprise so manifestly beneficial to the interests of the Ottoman Empire, as to those of all other nations, cannot fail to be given, now that science has pronounced in its favour, and that sufficient capital is ready to carry it out.

" Admitting, then, that the Sultan, to begin with, is with one accord with his vassal, the Viceroy will place himself in a very favourable posture as regards Europe, if, in order to prevent any further difficultie s between the friendly Powers themselves or with Egypt, he asks the former to designate plenipotentiaries to

Constantinople for the purpose of regulating by means of a convention the perpetual neutrality of the passage through the Suez Canal, the principle of which, in so far as regards the Ottoman Empire, is already set forth in Clause 14 of the Act of Concession.

"In this way the internal question of the execution of the canal is kept separate, as it should be, from the external question of neutrality. The prerogatives of the territorial sovereignty remain intact, and the Ottoman Empire, assuming for the first time since the conclusion of peace the influential position which it has a right to occupy in a negotiation of public European law, satisfies the political and commercial interests of all the Powers, while it at the same time obtains, by their accession, a fresh guarantee of its integrity and independence.

"The Viceroy of Egypt, who has so faithfully served his Suzerain during the war, will have rendered, by his conduct in regard to a work of peace, a not less signal service, and thus will be fulfilled the prediction of Napoleon I. at the beginning of the century, that the execution of the canal from sea to sea would contribute to the glory and to the maintenance of the Ottoman Empire."

To Mr. Richard Cobden, M.P., London.

"CAIRO, *November* 22, 1856.

"Five years ago I informed you that the Viceroy of Egypt had determined to make a maritime canal through the Isthmus of Suez. It was a time when our two countries were united in war that I called your attention to a work of peace, of progress, and of civilisation. I invoked the aid of your enlightened influence in case some members of the aristocracy, masters of the course of affairs, should be blinded by inveterate prejudice and narrow sentiments of exclusiveness and rivalry to oppose the execution of a work of universal interest. Since then I have had the opportunity of discussing with you personally this interesting question. I told you that the resistance which I had foreseen had become a fact, but that in a country of free discussion like yours I considered that the main point was to enlighten public opinion, still but partially informed, and to clearly demonstrate the feasibility of the undertaking.

" This appeal to public opinion has been made. The most experienced engineers in Europe have visited the scene of operations, and have published their final report; capital has been subscribed to commence the work; the Viceroy himself has placed himself at the head of an enterprise which has the unanimous and hearty support of the press in Europe and America; and, finally, the adhesion of the different governments has kept pace with public opinion.

" One only difficulty has arisen, and that is the opposition of your Government, which, through the influence of its ambassador at Constantinople, has succeeded in obtaining the suspension of the formality

of the ratification asked for from the Porte by the
Viceroy in favour of a concession which he had legally
granted. In so just a cause I shall not be at a loss
for means to overcome this obstacle, against which I
have deemed it useless so far to struggle, because it
could not up to the present hinder the march of the
enterprise, and because all the preliminary investigations not being yet terminated, we were not yet in a
position to put the work practically into execution.

" In a short time the situation will have altered, and
in order to avoid if possible the drawbacks which would
result from a conflict, you will not be surprised to
find me still appealing to the good sense of the public.
My opinion is that all this business is calculated to
revive ill-feeling between France and England,
whereas it would be desirable to see the sincere union
of the two peoples succeeding the uncertain and
already weakened union of the two Governments. If,
upon the one hand, the country which has proclaimed
freedom of trade is so inconsistent with itself in a
question relating to the freedom of commercial trans-
actions, and if, upon the other hand, France becomes
persuaded that her former ally has two weights and
measures, it is clear that all the efforts of sensible
men will one day fail before a fresh explosion of the
ancient prejudices which have so long separated the
two nations.

" For how can it be imagined that people on the
Continent will believe in the sincerity of England, in

her zeal for universal progress, civilisation, and public
wealth, if it is seen that England, where public opinion
reigns supreme, allows her Government to continue
its incredible opposition to the Suez Canal, a private
enterprise, in the origin, constitution, and object of
which there is nothing to awaken any suspicion of
political rivalry? How can the apostles of free trade
and open competition propagate their doctrines when
the two leading members of the Cabinet, who recently
figured in their ranks, will not agree, through fear or
horror of competition, to the suppression of a narrow
neck of land which divides the two most opulent
of seas, and stands as a feeble barrier against all the
navies of the globe?

" One of your greatest ministers spoke as follows at
a sitting of the House of Commons, when a vote was
taken which has reflected so much glory upon him :—

" ' You have to choose between progress towards
liberty and a return to prohibition ; you have to choose
the motto in which the commercial policy of England
will be made manifest. Will it be : " Advance," or " Go
back " ? Which of the two words best suits this great
Empire ? Consider our position, the advantages which
Providence and nature have conferred upon us, and
the destiny which awaits us. We are situated at the
extremity of Western Europe, the principal ring, as
it were, which connects the Old World and the New.
The discoveries of science and the improvements in
navigation have already brought us within ten days

of St. Petersburg, and will soon bring us within ten
days of New York.   A larger stretch of coast, in proportion to our population and the superficies of our
soil, than is possessed by any other nation, ensures
for us maritime superiority and force.   Coal and iron,
the nerves of industry, give our manufacturers great
advantages over those of our rivals.   Our capital
exceeds that of which they dispose.   In inventive
power, in skill, and in energy we are second to none.
Our national character, the free institutions under
which we live, our liberty of thought and action, an
unfettered press, which rapidly spreads abroad discoveries and progress—all this places us at the head
of the nations which naturally develop by the free
exchange of their produce.   Is this the country to
dread trade, a country which can only prosper in the
artificial atmosphere of prohibition ?   Choose your
motto, "Advance," or "Go back " ? '

 " It was on March 27th, 1846, that Sir Robert Peel
made this speech, and when the Free Trade Bill
subsequently came back to the House of Commons,
having been passed by the House of Lords, he added :—

 " 'The name which must and will be placed at the
head of this achievement is not that of the noble lord
(Lord John Russell) who leads the party which has
voted with us, nor my name ; it is the name of one
who, by the purity of his motives and his indefatig-
able energy, has appealed to the reason of us all, and
who has compelled us to listen to him by force of an

eloquence all the more admirable because it was without pretence or ornament—it is the name of Richard Cobden.'

"To you it now belongs, armed with the experience of the last ten years of progress and prosperity, which have been assured to the British Empire by the triumph of your system, to maintain the principle of free competition abandoned by some of your former companions-in-arms, and to place before your fellow-countrymen once more the alternative of progress or retrogression. The force of your convictions and of public opinion will not fail to secure for you a success in which the honour and profit of England are alike involved.

"I have no doubt that the question will be brought before Parliament. If you will consent to take up its defence, my friend and fellow-worker M. Barthélemy St. Hilaire, a Member of the Institute, will give you all the information which you may desire. He will shape his course as you may advise, and, at the time you may think proper, he will put himself in concert with you and other friends. I have asked him to hand you this letter, and to establish with you the relations which you will, I am sure, be happy to form with so distinguished and honourable a man."

END OF VOL. I.

PRINTED BY J. S. VIRTUE AND CO., LIMITED, CITY ROAD, LONDON.

Printed in the United States
By Bookmasters